Agriculture in the City's Countryside

Agriculture and urbanization are often considered mutually antagonistic forms of land-use. But in an increasingly urban-orientated world, agriculture depends on city markets and often has to operate within the complex mosaic of land-uses and activities of the urban fringe.

C.R. Bryant and T.R.R. Johnston provide a broad analysis of the nature, structure, and relationships of farming in the countryside surrounding cities. From the perspective of resource management and land-use planning, they explore the biophysical and human dimensions in the organization of periurban agriculture and identify four main elements around which they present their analysis: the resource base, the market, the farmer, and the role of government.

Case study material from both developed and developing worlds demonstrates how conflicts and problems can be resolved to the benefit of farmer and city-dweller. The authors also use systems theory to put forward a synthesis for understanding how modern agriculture functions under urban influence and how these trends are likely to develop in the contexts of conservation, economics factors, and technological change.

C.R. Bryant is a member of the department of Geography, Université de Montréal.

T.R.R. Johnston is a member of the department of Geography, University of Lethbridge.

Agriculture in the City's Countryside

Christopher R. Bryant
and
Thomas R.R. Johnston

University of Toronto Press
Toronto. Buffalo

© Christopher R. Bryant and Thomas R. R. Johnston 1992

First published in North America by
University of Toronto Press 1992

ISBN 0-8020-2842-X (hb)
   0-8020-7723-4 (pbk)

All rights reserved. No part of this publication may be reproduced, stored in a retrieval system, or transmitted by any other means without the prior permission of the copyright holder. Please direct all enquiries to the publishers.

Canadian Cataloguing in Publication Data

Bryant, Christopher Rex
   Agriculture in the City's Countryside

   Includes bibliographical references and index.
   ISBN 0-8020-2842-X (bound)   ISBN 0-8020-7723-4 (pbk.)

   1. Agricultural geography. 2. Agriculture - Economic aspects. 3. Land use, Rural. I. Johnston, Tom. II. Title.

   S494.5.G46B7 1992   338.1   C91-095557-3

Typeset by Communitype Communications Ltd
Printed and bound in Great Britain by Biddles Ltd, Guildford and Kings Lynn

Contents

List of figures	vi
List of tables	viii
Preface	xi

1.	Agriculture in an urban world	1
2.	The resource base: the underpinnings	24
3.	The market: systems and structures of exchange	55
4.	The farm entrepreneur and the farm	100
5.	The government: intervener in the enabling environment	137
6.	The nature of agriculture in the City's Countryside: a synthesis	190
7.	Retrospect and prospect	199

Bibliography	206
Index	227

List of figures

Figure 1.1	Post-industrial society and its driving forces	4
Figure 1.2	The form of the regional city	7
Figure 1.3	Forces affecting agriculture in the City's Countryside	11
Figure 1.4	Post-industrial society and agriculture	17
Figure 2.1	The spatial coincidence of population and highly productive agricultural land in Australia, 1976	26
Figure 2.2	The removal of agricultural land from agricultural production	29
Figure 2.3	Schematic representation of the relationship between the quality and volume of agricultural land	37
Figure 2.4	Relationship between provincial land capability for agriculture and conversion of prime land, Canada, 1966–86	41
Figure 3.1	Major paths in the marketing of farm produce	62
Figure 3.2	Pick-your-own calendar: an example from a large PYO north-west of Paris	65
Figure 3.3	Diffusion of selected items of farm machinery in the metropolitan (MR) and non-metropolitan regions (NMR) or Canada, 1941 to 1976	77
Figure 3.4	The Thunian formulation of agricultural land use location	81
Figure 3.5	Agricultural land value curves around an urbanising area	85
Figure 3.6	Different forms of farm fragmentation in a near-urban environment	88
Figure 4.1	Basic relationships in the farm operating system	101
Figure 4.2	Farm decision system and the external environment	104

LIST OF FIGURES

Figure 4.3	Conceptual framework for farm adjustment in near-urban areas	117
Figure 4.4	Effects of additional income benefits from near-urban locations on the setting aside of farmland	121
Figure 4.5	Policy and strategy options at the farm level in the City's Countryside: key examples	124
Figure 4.6	Farmers' evaluations of an agricultural future for their land and potential urban development	132
Figure 5.1	Landownership as a bundle of rights	142
Figure 5.2	Images of the countryside and the realities of modern farm production	185

List of tables

Table 1.1	Functions of agricultural land: examples	21
Table 2.1	Agricultural capability of land around Canada's 23 largest cities	25
Table 2.2	Examples of estimates of the conversion of agricultural land to urban uses in selected countries	27
Table 2.3	Considerations in evaluation of the significance of farmland conversion to urban uses	36
Table 2.4	Distribution of agricultural land in Canada by agricultural land capability and by province	38
Table 2.5	The distribution of land by agricultural capability class within an 80 km (50 mile) radius of Canada's 23 Census Metropolitan Areas	39
Table 2.6	The urbanisation of rural land in Canada, 1966–76, by agricultural land capability	40
Table 2.7	Future urban growth scenarios for Ontario used by Smit and Cocklin (1981)	43
Table 2.8	Land requirements from agriculture for different urban growth scenarios, Ontario, 1976–2001	44
Table 2.9	Estimated rural-to-urban land conversion (1976–2001) as a percentage of total land in Ontario by agricultural capability	44
Table 2.10	Scenarios used by Smit et al. (1983) to assess the implications of future urban expansion on food production	45
Table 2.11	Results of Stonyer's technique for expressing the implications of farmland urbanisation in terms of lost export earnings	46
Table 3.1	Agricultural production changes in selected US metropolitan counties, 1949–82	58

LIST OF TABLES

Table 3.2	Changing patterns of agricultural intensity, Southern California, 1950–80	59
Table 3.3	Farms reporting selected agricultural produce, Sydney Statistical Division, 1981–2	59
Table 3.4	Produce types in direct selling in Connecticut, 1989	66
Table 4.1	Dominant values in farming	106
Table 4.2	Values held by West Midlands hop farmers	107
Table 4.3	Urban-based problems experienced by farmers near Toronto	122
Table 4.4	Farm operator adaptive behaviours near Toronto	126
Table 4.5	Farm diversification in the Montreal south bank and Toronto SW areas	127
Table 5.1	Private and public interests in agricultural land	140
Table 5.2	Rationale for public sector involvement in agricultural land and the sustainability of agriculture	153
Table 5.3	Approaches and perspectives to agricultural land resource issues in the City's Countryside	160
Table 5.4	Techniques of public intervention in relation to the agricultural land base	166
Table 5.5	Evaluation of the rationale for Ontario's Food Land Guidelines	178
Table 5.6	Examples of strategies for goal and type of agricultural area combinations	187
Table 6.1	Key characteristics of agriculture in the City's Countryside compared with areas beyond	191

Preface

Despite the tremendous changes that have occurred in the twentieth century in Western society in terms of economic development and urbanisation, agriculture remains a critical element in our daily lives. Agriculture is one of the cornerstones of our life support system, and for that reason alone is always a critical component of the search for a sustainable development strategy for any society. Agriculture also supports landscapes which have environmental and cultural values embedded in them, and it provides the most frequent point of contact with 'nature', albeit greatly modified by human activities, for the vast majority of the urban population. Finally, agriculture still occupies a major part of the settled portion of the Western world, even though the population that manages and lives by it has dwindled appreciably.

It is because of the significant concentration of populations in urban regions and the multiple demands made by this population on the nearby rural areas that the agricultural activity close to urban areas has received so much attention from geographers, planners and politicians in the second half of the twentieth century. Concerns about loss of agricultural production capacity, for instance, have surfaced in one country after another as urban land uses have expanded, and literally dozens of programmes of farmland conservation have been developed and implemented. At a time when agricultural overproduction has been a major preoccupation for the agricultural economies of many Western nations, such concerns seem misplaced; yet continued problems regarding famine and undernourishment in many parts of the world and recent worrisome signals linking the search for continual increases in agricultural production efficiency to deterioration in the productive capacity of the agricultural land resource in many developed countries are providing more grist for the agricultural land conservation mill. And these concerns have been become more widespread as the concept of sustainable development, however difficult it may be to define, has caught the eyes and imagination of more and more people.

Furthermore, other interests in or values attached to the biophysical and built environment in the City's Countryside have become increasingly significant for the populations of urban regions, partly as a reaction to the excesses and complexities of the modern urbanised society in which they live. All this bears witness to a complex set of transformation processes affecting agriculture in the City's Countryside and of the interests that exist in the agricultural production system there.

Agriculture in the City's Countryside reflects a synthesis of our own teaching and research. It owes a great deal to our late colleague and friend, Lorne Russwurm, who co-authored *The City's Countryside*, published by Longman in 1982, with Bryant and another colleague, Sandy McLellan. That book presented a synthesis of a wide range of processes of change affecting different land use activities in the City's Countryside. *Agriculture in the City's Countryside* arises logically and conceptually from the discussion of agriculture in *The City's Countryside* as well as from the joint research of Bryant and Johnston since 1985.

We hope that this book will prove stimulating both to students and professionals in the field of rural land use and resource planning and management by presenting a series of conceptual frameworks with which the complexities affecting modern agriculture near cities can be teased out. We hope as well that this will provide a critical perspective on the various programmes and plans that various public agencies have elaborated to tackle the 'problems' faced by agricultural land and communities in the City's Countryside.

Two basic themes provide the basic conceptual framework: a systems perspective and an 'interests-in-land' perspective. The systems perspective provides a unified structure for investigating agriculture in the City's Countryside, both in terms of the geographic and societal context in which modern agriculture functions and of the functioning of agriculture itself. Four fundamental dimensions of the system in which agriculture functions are singled out: the resource base, the 'market', the farmer and government.

The biophysical resource base is singled out partly because of the heated debates concerning the effects of different processes of change on the agricultural resource base. Even though the sustainable development concept embraces much more than the physical environment, the 'natural environment' is none the less still a central part of the concept. The 'market' dimension is seen as providing the principal integrating mechanisms between resources, producers and consumers and, as argued later, provides the basis for maintaining the vitality of agriculture in the City's Countryside within a sustainable development context. The farmer as entrepreneur and manager of the farm unit is presented as the fundamental unit of analysis and decision-making which must be understood in order to comprehend changing agricultural structures and land uses in the City's Countryside. It is also important to focus on the farmer and the farm unit in order to understand the limitations and special challenges that face public intervention in the domain of land use planning and management in the Western world.

Interests in land in the Western world have become increasingly complex, as more and more interests have been established in the agricultural resource base besides the obvious one of private economic gain from engaging in agricultural production. These interests include the possibilities of reaping speculative gains in the capital value of the land, land as a support for the family farm system, land for food production for present and future generations and land as a support of amenity. These different interests are not uniformly held, either between Western countries or even within the same country. The interests in land perspective is thus a recurring theme that provides a framework for investigating the diversity of private and public responses to the changes that have been transforming agriculture in the City's Countryside.

Finally, underlying these two perspectives are two other recurrent themes: individual versus collective interests in land, and technological change. The first theme reflects the ever-present tension between individual rights in landownership and the use of the resources in land for the benefit of the broader collective interest. Land use planning and management can be thought of as a continual process of trying to resolve these tensions. The second theme, technological change, has been central to the process of transformation of the agricultural production system and in the creation of an increasingly interdependent national and international food production system. Technological change especially through the private car, has also been one of the driving-forces behind the development of new settlement forms in urban regions and, therefore, in the City's Countryside in the twentieth century.

The systems perspective is developed in Chapter 1, while the main components singled out for emphasis – the resource base, the market, the farmer and government – are treated respectively in Chapters 2, 3, 4 and 5. A synthesis of the nature of agriculture in the City's Countryside is developed in Chapter 6. Throughout the first six chapters, illustrations are given from our own empirical research as well as that of other researchers in the Western world, but the emphasis is on frameworks. In the final chapter, we speculate on the future of agriculture and on the role of government in the dynamics of agricultural change in the City's Countryside.

We should like to acknowledge with thanks the various pieces of research and literature provided by Peter Crabb, School of Earth Sciences, Macquarie University, Australia, as well as the numerous colleagues from Europe, the United Kingdom, North America and New Zealand who have discussed with us many of the ideas developed in this book. We would also like to thank Karen Puklowski of the Department of Geography, Massey University, and Guy Frumignac of the Département de Géographie, Université de Montréal, for their cartographic assistance.

Christopher R. Bryant
Université de Montréal, Québec, Canada
Thomas R.R. Johnston
University of Lethbridge, Alberta, Canada
July 1991

1
Agriculture in an urban world

Rural and urban worlds

Agriculture, the purposeful tending of crops and livestock, continues to occupy a special role in our lives. Despite the rapid decline in farm numbers, farm population and even farm area in most Western countries during the second half of the twentieth century, agriculture accounts for one of the cornerstones in our life support system – our daily sustenance. Agriculture also supports other values that have become increasingly important in our urbanised society: it supports landscapes in which part of our historic and cultural heritage is embedded; it modifies, contains and is supported by the biophysical environment; and it is associated with open spaces around our cities, contributing to the provision of outdoor recreational opportunities for the urban population.

Agriculture occupies a curiously paradoxical situation with respect to the natural environment. Despite the declining role of land in farm input structure in many regions, the increasing importance of capital for inputs of non-farm origins and the emergence of highly specialised 'factory' farms in some agricultural sectors and regions, agriculture continues to retain a strong link with the biophysical environment. The soil resource, for most agricultural enterprises, still represents a major input into the production process, either directly (on-farm production) or indirectly (e.g. livestock feed purchased off the farm). At the same time, agriculture has a strong link with the city and, more generally, the urban–industrial complex. There are strong historical links between agricultural development and industrialisation and urbanisation, with agricultural surpluses feeding the growing cities during the Industrial Revolution and providing important parts both of the labour and capital required in these processes (Grigg, 1984; Pautard, 1965). The urban–industrial complex has also played a significant role in fuelling mechanisation and motorisation in agriculture, and strengthening the interdependencies between the agricultural and non-farm sectors of our economic systems.

On a more regional scale, agriculture near cities has often responded to the nearby market by producing various specialty agricultural enterprises for the urban market, e.g. vegetables, fruit, poultry and milk. Furthermore, the agricultural land resource supports a good part of the rural landscape around cities and therefore contributes to the open space function that is valued by the urban populations in the pursuit of their recreational needs. It is because of these demands upon agriculture, as well as other demands that have an effect upon the numerous activities and functions carried out in the countryside around cities, that the term 'City's Countryside' was coined (Bryant et al., 1982). We should remember, however, that farming in the City's Countryside also serves markets beyond the City's Countryside, and we shall have reason to return to this point many times.

Agriculture in the City's Countryside is in a paradoxical situation. While it derives value from its links with the urban-industrial complex, it is also undergoing stresses that emanate from the urban-industrial complex (as well as from other sources). This is perhaps not surprising because any system or structure that evolves is likely to experience stress. The apparently contradictory situation with respect to the urban-industrial complex can also be partially explained by the variety of different values, interests in and images associated with farming.

The many differing values, interests in and images associated with farming will be a recurring theme in our discussion. They are held by different people and groups such as farmers, policy-makers and the public at large, and with respect to different aspects of farming such as the land resource, the farming population and the products arising out of the production process. Images are important because they reflect assumptions about reality and the values we associate with farming; they therefore influence research directions, public policies towards agriculture and individuals' decision-making behaviour. Imagine the response of a policy-maker whose dominant view of farming is that of an idyllic pastoral activity or who assumes that agriculture should be dominated by the 'family farm', and contrast this with someone who views agriculture as necessarily moving further towards an efficient, industrialised production machine. Similarly, different people and groups have different images of agricultural change: there is a world of difference between the image of agricultural change leading to the collapse, demise and degeneration of the agricultural production system and that of an agricultural structure adapting and being transformed (Bryant, 1989a).

All of these different images, be they of the structure itself or its changes, are but partial representations of agricultural activities, especially of those in the City's Countryside. Agriculture is a heterogeneous activity and agricultural change is complex, without a single evolutionary path. Unfortunately, the images of agriculture which are but partial representations of reality have helped create a conventional wisdom that has influenced research directions and policy. This has arisen partly at least because of a lack of appreciation of where agriculture sits in relation to the broader socio-economic system. Therefore, in the remainder of this first chapter, we consider this relationship, especially as it is manifest in the specific settlement context of the City's Countryside and the nature of the agricultural production system; finally, we consider the implications of this broader view of agriculture for understanding the different images, values and interests in agriculture and their links with the management and planning activities that touch agricultural activities.

Agriculture in post-industrial society

To be truly appreciated, agriculture must be seen as a complex system, which in turn is part of a broader macro system, in which a variety of transformations lead to different geographic patterns of response manifest on different geographic scales. To understand agriculture, then, it can be argued that it is first necessary to comprehend this broader functioning system, not just nationally but globally too (cf. Marsden et al., 1990). This converges with the increasing pleas for the effective integration of a political economy perspective into the analysis of agriculture (see, e.g., Marsden, 1986b). We first develop the idea of different forces operating on different scales which can impinge upon agriculture, and then we discuss the context of the City's Countryside, the particular geographic focus of this book, before dealing specifically with the socio-economic structure of agricultural production and its links with the broader system.

During the middle part of the twentieth century, attention began to be directed towards the evolution of a different form of social and economic structure, superseding the industrial society – the post-industrial society (Bell, 1973). This evolving system has been characterised, among other things, by the rapid development of a group of heterogeneous service-oriented sectors, the growing importance of information as capital, an increasingly open economic system, facilitated by the rise of information as a key factor of production and the massive advances in communication technology, and the growth of new consumer 'needs', for instance in terms of leisure. While the industrial production model has yet to be replaced, the social and economic systems of Western nations in the last quarter of the twentieth century are a far cry from their antecedents of fifty years earlier. Continual change seems to have become accepted as a fact of life, and it is fashionable to speak about macro or megatrends and how they are affecting our lives (Naisbitt, 1982).

How do these macro trends or transformation processes that are both creating and are characteristic of post-industrial society affect the agricultural system? To provide some insight into this, it is useful to think first, of human activities as functioning within systems of interaction or exchange which operate on different geographic scales, and second, of how the various transformations or macro trends have both worked their way through these different systems and modified the boundaries of the systems of exchange. It is these exchanges that provide us with an important conceptual tool with which to analyse agriculture in the City's Countryside in its broader context. Munton et al. (1988) claim this is essential if we are to understand farm change in this particular type of environment.

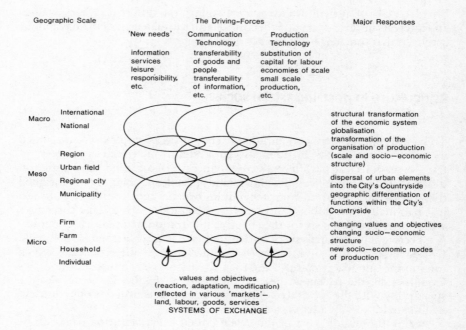

Figure 1.1 Post-industrial society and its driving forces

First, there are a whole series of different scales of geographic analysis (Figure 1.1), ranging from the macro scale (international and national), through the meso scale (broad region, urban field, regional city, municipality) to the micro scale (firm, farm, household, individual). Socio-economic systems function through interaction; while part of this interaction can be seen as functioning between 'units' on a given geographic scale of analysis, e.g. between individuals in a community, between urban regions in a national system of cities, or between nations at the international scale, it is of at least equal importance to appreciate the systems of exchange (goods, services, people, information, ideas, capital) that tie units from different scales together. Thus, the individual farm unit is tied into the national system of farm production, the national and international farm supply industries, and the international market-place for farm commodities. Certainly, these systems of exchange can be more or less localised depending upon the nature of the production system: contrast wheat production for the world market with fruit and vegetable production for the regional urban market. However, the key point is that to understand the nature of change in agricultural production, it is important to realise that there are different systems of exchange which the different farm units, even in a given area, can participate in and belong to, and that the boundaries and configurations of these systems of exchange can themselves change over time.

The exchanges can link decision-making units on one scale of geographic analysis with those on other scales or with aggregations of decision-making units such as other competing producer regions. The exchanges can involve people through, for instance, the daily journey-to-work movements, seasonal migration patterns and permanent migration patterns. They can also involve goods and commodities (e.g. farm produce, farm inputs), information (e.g. news, information on relative prices and competitors' strategies), and money, capital and ideas.

It is important to appreciate that the boundaries placed on the systems of exchange within which particular units such as farms function, as well as the direction and magnitude of exchanges, can be altered by various transformations. The origin of the transformations may appear to come from specific geographic scales of analysis. For instance, commodity price changes resulting from supply and demand relationships in the international marketplace, may reflect aggregate supply and demand relationships, political decisions and international trading structures. However, such changes always work their way eventually through all levels of the particular system of exchange, creating potentially different effects on each scale level.

What are these broad transformations that generate and reflect changing signals to those involved in agricultural production? Three sets of forces or processes are suggested as characterising post-industrial society: the development of new 'needs' in society, changing communication technology and changing production technology (Bryant, 1988; Bryant and Coppack, 1991). These have influenced all economic activities to greater or lesser degrees, and each potentially alters the values associated with different exchanges as well as the resources and production associated with different nodes in the systems of exchange. Some brief comments are offered next on the links between these forces or processes and agricultural production.

For agriculture, the reduced propensity to consume more food products as incomes increase has placed certain limits on the domestic market for food products but there has been an increase in the demand for non-food items such as horticultural and nursery production. Increasing demands for produce that is healthier and freer of contamination from herbicides and insecticides is altering the demand for agriculture produce (Blunden and Curry, 1988; Marsden et al., 1990). Growing demand for recreation opportunities has provided some farmers with the potential to add recreational enterprises to their farm units. Finally, the increasing need for personal fulfilment can be partly linked to the decentralisation of responsibility (and perhaps to the mounting pleas for deregulation of government attempts to control agricultural production in the 1980s), while the emphasis on the non-economic values and consequences of agricultural production have helped to emphasise the existence and to spawn the development of different socio-economic modes of organisation of agricultural production.

Communication technology has already played an important role in the transformation of agricultural production on world and national scales. Part of this has been rooted in the processes of industrial society, leading to greater ease of transportation of agricultural commodities and products and enlarging the feasible supply area for foodstuffs for major markets.

However, another part has been more recent and deals with communication and processing of information, and is firmly embedded in post-industrial society. Its effects upon agricultural production and organisation are more subtle, yet of significant magnitude. It can contribute to the reorganisation of management systems at the individual farm level; it increases the openness of the system; and it can substantially alter modes of marketing agricultural produce. A particularly powerful example of the role of communication technology is seen in the functioning of the flower auction market at Aalsmeer in the Netherlands near Amsterdam. Here, information technology permits the efficient sale of over 13 million flowers and 1 million pot plants daily, and the transportation technology of truck, air and refrigeration ensures that the produce is on sale at its various destinations throughout Europe and the world on the same day (VBA, 1989).

Change in production technology also has its roots in industrial society, and much of the pattern of agricultural change which we have come to accept as 'normal' represents the integration of the industrial model of production into the agricultural sector. This pattern includes mechanisation, substitution of capital for labour, the increasing reliance of farm production on inputs produced in the non-farm sector, increasing farm business size in order to benefit from economies of scale, and adopting increasingly sophisticated management and decision-making structures (Marsden et al., 1986a). Other production technology changes are more linked to post-industrial society, for example biotechnology with its enormous demand for knowledge as a factor of production.

These broad forces or processes obviously overlap and operate simultaneously; the changes in values they imply or lead to work their way through all of the various geographic scales of analysis via the different systems of exchange. However, we should not think of these macro forces or processes as only working their way down through the geographic scales of analysis; there are upward linkages too as individuals respond, sometimes reactively it is true, but sometimes proactively as well. Collectively, these individual decisions also help shape the system and its dynamic evolution. We shall return to the role of the individual farmer in Chapter 4.

Settlement organisation in post-industrial society

The broad forces and processes associated with post-industrial society have given rise on the meso scale to a changing settlement structure. It is this regional or meso scale organisation of the settlement system which is one of the dominant characteristics of post-industrial society. This structure, the regional city (Bryant and Coppack, 1991; Bryant et al., 1982), is a functioning social and economic system (Orhon, 1982), involving both concentrated urban core areas as well as more open, dispersed countryside (see Figure 1.2). Some have called this whole structure the 'urban field' (Friedman, 1973; Hodge, 1974), but we prefer to reserve this term for the more dispersed part of the regional city (Bryant et al., 1982; Coppack et al.,

1988). It is this dispersed part of the regional city that is also termed the 'City's Countryside' to reflect the special relationships between the rural and urban parts of the regional city. It has much in common with the idea of the 'accessible countryside' (Blunden and Curry, 1988).

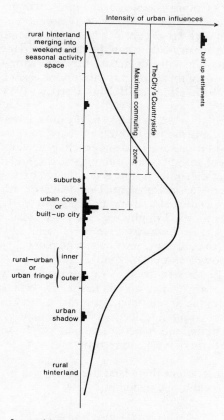

Source: Adapted from Russwurm, 1975, 151

Figure 1.2 The form of the regional city

The regional city can be seen as a series of overlapping and more or less integrated systems of exchange that have developed between various functions, both those thought of as traditionally urban (e.g. factory work, service employment, residence, certain types of leisure) and more rural ones (e.g. outdoor recreation, certain agricultural activities). Urban elements have been spread over larger and larger areas as the boundaries of the systems of exchange of which they are a part have been modified by improved communication and transportation. The changing geographic patterns within the regional city result both from redistribution of some urban elements as well as differential patterns of new growth within this regional settlement structure.

Inevitably, these changes are reflected in changing form (e.g. the location of different activities and land uses), function (i.e. the *raison d'être* of different uses of the land) and structure (i.e. interrelationships). In terms of land use, the pattern in the City's Countryside has been represented in schematic form as a series of land use zones, characterised by decreasing intensity of the urban elements as the concentrated urban core area is left: the inner urban fringe, the outer urban fringe, the urban shadow and the rural hinterland (Bryant et al., 1982) (see Figure 1.2). We will not debate here the issue of defining the rural–urban fringe or any of these zones as this is done elsewhere (see Bryant et al., 1982; Firey, 1946; Hind-Smith and Gertler, 1963; Pryor, 1968; Russwurm, 1977).

Briefly, the 'inner fringe' is characterised by the most intense development of urban elements beyond the urban built-up core; here is where we can expect to find rural and agricultural land under active development pressures, and the major portion of any speculative activity in the land market (Bryant, 1982). The landscape of the 'outer fringe' is dominated by rural activities and uses, but there is likely to be substantial evidence of urban infiltration. Depending upon the specific cultural and policy framework with respect to land use development, this might be seen by scattered residential developments along major highways, scattered and isolated residential development, clusters of estate or country residential development or simply the expansion of the towns and villages in the area.

The outer fringe gives way to the 'urban shadow' where urban or metropolitan influences are evident only in terms of a certain amount of land in non-farm ownership, the development of some country estates, and a less intense development of country residential development. This zone finally merges into the rural hinterland, where there may still be some evidence of urban elements in the form of weekend retreats, and cottage or second home developments.

The specific configuration of these land use zones will be influenced by a host of unique, local conditions, including:

1. physical geography (e.g. the rapid reduction in urban pressures southeast of Montreal once the Richelieu River is crossed (INRS, 1973));
2. land tenure patterns (contrast the more distinct edges of the cities in the Great Plains of the United States and Canada because of the distinctive large blocks of land under single original ownership with the tremendously fragmented patterns along the edges of many West European cities such as Paris);
3. cultural patterns and attitudes (e.g. the resistance of the Mennonite farmers in the rapidly urbanising area of Kitchener-Waterloo in Southern Ontario, Canada, to sell their land for development);
4. political factors (e.g. the fragmented local municipal structure in the urban fringes of many cities can influence when and where development takes place); and

5. planning policy and land use control regulations (contrast the zones around most British cities which have evolved under a long-established town planning tradition with those around many American cities, a country where planning agencies have had less control over development).

What of the role of agriculture in this situation? It is common to see agriculture as simply reacting to the urban development pressures, both of the accretionary and the more scattered types, and just giving way. However, agriculture is not a homogeneous activity and can vary even within a given regional city in terms of its resource base, the nature of the land tenure system, the cadastral structure within which it functions, and its overall viability. Therefore, at the very least, certain types of agricultural structure can be expected to provide short-term rigidities in the supply of land for development, while its underlying structure (especially the field and property structure upon which it is based) can influence the nature and form of the development (contrast the mixed, low-density patterns of urban development in the former vegetable and fruit-growing areas to the north-west of Paris, France, with the large-scale residential complexes and even New Towns that have evolved since the mid-1960s on the plateau areas west and south-west of the same agglomeration).

It is well to remember that the pattern represented in Figure 1.2 is an idealised one, which emphasises development and urban influences of varying intensities. In reality, the form of the City's Countryside is more likely to be characterised by discontinuities in the arrangement of uses and changes in land use. We shall return to this idea of a mosaic of land use zones, especially in relation to agriculture, several times in our discussion. The other point to emphasise is that the form of land use represented in Figure 1.2 really only represents one particular stage in the development of the City's Countryside in the regional city. The evolution of the regional city can be thought of as having occured through a number of stages (Bauer and Roux, 1976; Bryant et al., 1982), moving from the early beginnings of accretionary urban growth around the urban core when much of the growth was being fed by immigration of population from rural areas, to stages when public and/or private transportation (the car) and continued growth pressures created the conditions for dispersal first of residential development and then of commercial and industrial activities into the City's Countryside, and finally to the stage of development of megalopolitan structures – of overlapping regional cities – where the density of regional cities allows this.

The land use patterns therefore vary over time; furthermore, not all regional cities have developed to the same degree, partly because growth pressures vary enormously between cities in any given national space economy. So we can expect the role of agriculture in the land use patterns to vary geographically, not only within a particular City's Countryside, but also between regional cities.

Furthermore, stresses are created in any dynamic structure, and the evolving regional city is no exception to this. The above discussion suggests that the stresses felt by agricultural activities can be expected to vary geographically as well as temporally. It is important to remember, however, before we focus on the links between the urban components of the City's Countryside and agriculture, that the regional city is a meso-scale manifestation of a whole series of processes and forces operating through a particular subset of systems of exchange. In the evolution of the regional city, stresses are inevitably created (i.e. different values and ideas come into conflict), but it is important to remember in our interpretation of the observed changes that the farm decision-making units are also tied into other systems of exchange through their production function.

In dealing with urban–agricultural relationships, it has been evidently too easy to put aside these other relationships. It has been common, and still is in many quarters, to attribute the stresses experienced by farmers in the City's Countryside during this post-industrial era to the evolution of the settlement system itself – the demand for land for various urban functions – and to the presumed divergence of values between the non-farm elements of the population and the farm population.

This focus on the negative impacts of certain metropolitan or urban-based forces was what characterised much of the literature on urban fringe farming until at least the mid-1970s (see commmentaries by Bryant (1976), Bryant (1984a), Bryant and Russwurm (1979) and Munton (1974)). There was also a tendency to gloss over the substantial regional and even subregional differences that exist in the regional environments in which agriculture and urban areas have evolved side by side (Bryant and Greaves, 1978).

From the mid-1970s, however, a view of urban fringe and urban field agriculture has evolved which is considerably more complex (see Figure 1.3). First, the framework emphasises the need to recognise the entire range of forces influencing agriculture, regardless of where the locus or geographic scale of the force or transformation appears to come from. A whole series of forces or changes in conditions have been termed non-urbanisation factors (or non-metropolitan factors). These include developments in technology, interregional and international competition and living standards or life-styles which are related to the broader transformation of our evolving post-industrial society. In addition, they include national and international (e.g. trade agreements) decisions that influence the environment within which agriculture functions.

The other set of decisions include urbanisation (or metropolitan) factors – in other words, a set of factors or forces that can be linked directly to the developing urban area adjacent to the City's Countryside. Obviously, this includes the demand for land for various urban and urban-related functions (e.g. residential, commercial and industrial, infrastructural and recreational development). This is the cause of the most spectacular, and the most obvious, impacts associated with urban development, such as land conversion, land speculation, farm fragmentation and abandonment of farmland; perhaps this is why they have received so much attention.

There are, however, other influences associated with the developing urban area, viz. the demand for produce and services that can be satisfied from the agricultural sector using agricultural resources and the demand for labour by the non-farm sectors of the economy (Bryant et al., 1982; Lockeretz, 1987).

AGRICULTURE IN AN URBAN WORLD

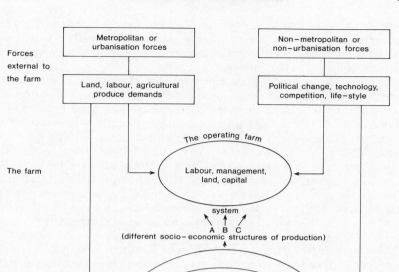

Figure 1.3 Forces affecting agriculture in the City's Countryside

There are noticeable concentrations of these demands in the main urban areas, but it is well to remember that they really represent meso-scale manifestations of the operation of the broader-scale processes associated with post-industrial society. We shall be returning to these various forces several times throughout our discussion.

The other focus, almost bias, of much of the conventional thinking about urban fringe and urban field agriculture has been the strongly negative image associated with changes in the farm structure related primarily to the demand for land for urban and urban-related functions. An important distinction has been made between direct and indirect impacts of urbanisation upon agriculture (Bryant, 1986a; Bryant and Russwurm, 1979; Johnston and Smit, 1985; Pacione, 1984), with the direct impacts referring to the actual removal of agricultural resources from agricultural production, and the indirect impacts referring to changes in the continuing farm structure. Most of the interpretation has been, and still often remains, dominated by consideration of negative impacts. Reality is more complex than this, however, and the more recent conceptual developments in this field have underlined the necessity to consider certain positive impacts too, e.g. the positive influences represented by the market opportunities for production based on agricultural resources, the potential ability to rent farmland relatively cheaply from non-farm interests, and the greater opportunities for off-farm employment often encountered in the City's Countryside.

Perhaps one of the best examples of how the negative interpretation associated with urban development impacts permeates much of our thinking comes from the various descriptive exercises that have been made to categorise the different zones in an urbanising environment based on the level of urban or non-farm population development (e.g. Krueger, 1978; Russwurm, 1970; Yeates, 1975, 1985). Based upon farm to non-farm population ratios, the areas in the City's Countryside have been classified along a continuum from rural, semi-rural, semi-urban and finally to urban. This has proved a useful device to describe the pattern of non-farm development, both at a municipal and submunicipal level, and the patterns of evolution of non-farm development over time (see especially Krueger's (1978) use of this to describe the pattern of evolution of non-farm development in the Niagara Fruit Belt of Southern Ontario in Canada). However, it has also been assumed frequently that resource use conflicts and 'problems' are related directly to this classification.

While it is reasonable to assume that there is indeed some relationship between level of urban development and stresses felt by farmers, it is not true that this is necessarily linked directly in any simple manner to structural problems in agriculture. Farmers certainly express having experienced 'problems' linked to proximity to the urban core and the distribution of their responses certainly reflects the relative levels of urban development (see, e.g. Bryant, 1981; Johnston, 1989); however, to make the automatic connection with level of resource use conflicts would require us to ignore the presence of viable agricultural production in the problem areas! We shall return to this later, but for now we can note that this apparent anomaly simply reflects either (1) that the 'problems' are present but are not really that significant for the continuing farm operation, and/or (2) other influences exist in the farmers' environment which can be capitalised on to cope with or circumvent the negative influences. Before assuming then that the continuum of urban pressures is directly linked to level of resource use and structural problems in farming, it would be well to undertake the necessary field work to verify such assumptions.

A final comment on the relationships between agriculture and urban areas concerns the modification of agricultural communities in the City's Countryside. Agriculture has undergone significant changes in the middle two quarters of the twentieth century in terms of labour input (see Chapter 3) which have had a profound impact on the viability and even existence of many agricultural communities. In the City's Countryside these same processes have been operating, but the observed changes in the agricultural communities are not as simple because of the influx of non-farm development in many areas, either concentrated into the smaller villages and towns embedded in the City's Countryside or as scattered non-farm residential development (the latter being particularly characteristic of North America). Here, population influxes may counterbalance the processes of decline caused by agricultural 'modernisation' and rationalisation and community populations have even increased in many situations, contributing to stability and even renewed prosperity for the service sectors present in the villages and towns (Coppack, 1985; Hodge, 1974). In his work on the impact of exurban settlers in the regions near Ottawa, Ontario, for example, McRae (1977) found that in the first two years after moving to the area, the exurbanites injected an average of $31,520 into the local economy over and above normal household expenditures.

Inevitably, changes occur in the community structure and the role of the agricultural community within it, leading eventually in many situations to a changeover in the local political power structure. This is seen by some people as further evidence of 'problems' for the agricultural community, because of the assumed differences between the agricultural and non-farm communities in terms of values. First, we have to note again that the evidence for significant differences in outlook in many parts of the City's Countryside seems to be ambiguous (Smit and Flaherty, 1981). Second, the juxtaposition of agricultural and non-farm communities, the fact that they often frequent the same institutions in the countryside (e.g. schools and churches) and the vast number of shared experiences through the various communication media inevitably mean that there is some integration of urban values, certainly in terms of expectations, into the farming community. There are some obvious differences, of course, which simply reflect differences in the nature of their respective economic activities and the different contact patterns that this entails (Walker, 1987). So there is some merging of the different groups (Glenn and Hill, 1977), but there are also differences reflecting the different activity base.

Third and finally, however, it is unwise to speak for too long about 'the' agricultural population or 'the' non-farm population in the City's Countryside without recognising the very significant differences that exist within each of these groups. In the non-farm population, we have to acknowledge that various groups of residents come into the countryside for different sets of reasons and motivations; within the farming population, the observation of heterogenity leads us to investigate the existence of different types of socio-economic organisation of agricultural production in the City's Countryside. The argument that we will build up to is for the recognition of an agricultural mosaic in the City's Countryside, comprising different socio-economic structures of production, and different farm enterprise systems responding in different ways to the variety of pressures faced by agriculture, urban-based and non-urban based alike.

A systems perspective on agriculture

The geographic focus of this book, the City's Countryside, is firmly set on the meso scale. Already, however, we have insisted several times on the need to recognise that individual farmers and their families in the City's Countryside are also caught up, to a greater or lesser degree in other systems of exchange which transcend the geographic scale of the City's Countryside. The various forces or transformations characteristic of our evolving society have been introduced above; they create signals that move through the various systems of exchange and create changes in the farmer's operating environment. How they are perceived and the strategies that the farmer and farm family then follow are influenced by personal factors and these in turn may also reflect differences in the particular mode of socio-economic organisation of agricultural production.

To unravel the complexities implied above, it is useful to adopt a systems perspective on agriculture and the organisation of production at the farm level. The individual farm operation can be seen as an operating system, combining enterprises, land, labour, capital and management into a functioning system (Bryant, 1984b; Olmstead, 1970). However, to appreciate the specific way in which these are combined, we have to understand the broader environment within which the farm exists, because each component of the individual farm operation is influenced to greater or lesser degrees by parameters that are set outside the farm operation. For example, the land component, at a particular time and place, is influenced by dominant patterns of land tenure and the legal environment in which farmland rental takes place as well as the property structure (e.g. degree of property fragmentation which in turn may influence field shape and sizes). Labour utilised on the farm may be influenced by institutional forces (e.g. legal arrangements regarding minimum wages and working conditions, as in France), as well as regional and macro conditions that influence the supply and demand for labour generally, and educational opportunities for agricultural training. The capital input is influenced by credit arrangements, interest rates, and the cost of inputs which in turn in a relatively open economic system is influenced by the demand from alternative uses of those same inputs.

Enterprises are influenced by ecological considerations (i.e. soil and climate conditions which influence the potential range of agricultural enterprises that can be produced in any given location), the market-place (i.e. supply and demand relationships), government subsidies (e.g. the agricultural price support system contained within the Common Agricultural Policy of the European Economic Community), and trade restrictions and agreements (e.g. the Free Trade agreement between the United States and Canada implemented in 1989). Finally, management and decision-making is influenced by values present in the broader society, by educational experience and opportunities and by information accessibility. In Chapter 4 we shall discuss whether the roles that farmers play as decision-makers in the City's Countryside are any different from farming beyond the City's Countryside when we consider the nature of innovative and adaptive decision-making behaviour.

The recognition of the importance of the broader environment in setting parameters which influence the way in which farms and agriculture are structured and function, led Malassis (1958) to argue that agriculture in the Western world had experienced a series of evolutionary changes in the course of its history, reflecting the broad changes that were occuring in the whole socio-economic system. For purposes of exposition, Malassis identified three 'pure' forms of socio-economic structure of agricultural production: subsistence, family farm and capitalistic (with a fourth one, co-operative, looming on the horizon in the 1950s). Each system was characterised, he argued, by distinctive social characteristics (i.e. specific relationships between the owners of the factors of production), economic characteristics (i.e. production for own consumption, production for both own consumption and for the market, and production for the market), technical characteristics (i.e. the relative role of tools and equipment in the production system), and financial characteristics (i.e. the relative use of credit in financing the farm). They were also distinguished by different motivating forces (i.e. production for family consumption, production to generate some cash, production to maximise profits).

How do these systems change, become modified or replace each other? The origins of the family farm system were linked by Malassis (1958) to the artisanal and guild organisations of the towns and cities of medieval Europe, while the seeds of the capitalistic system can be traced to the urban-industrial complex (Grigg, 1984; Malassis, 1958). The thesis is that as one system began to attain dominance in the urban context, it influenced the adjacent rural areas. Pautard (1965) argued convincingly that this meant that in nineteenth-century France, agriculture in the rural areas in the Paris basin and adjacent to the industrialising areas in the north and north-east of the country became transformed faster than elsewhere. Values associated with the rapidly developing capitalistic system in the urban agglomerations in these regions were transmitted and integrated into the rural work-force, with the resulting emigration of agricultural labour and farm operators. It then paved the way for farm consolidation and increases in farm size and farm mechanisation, and later motorisation, in these same regions. In the process of the spread of some of the attributes of the driving system in the overall socio-economic system, the earlier systems became transformed.

However, not all farms or regions are transformed to the same degree, and Pautard argued that this varied regionally as a function of the degree of interaction between the urban-industrial complex and the rural areas. What this means is that at any given time and in any given geographic space, it is likely that the farms there will represent a range of different socio-economic modes of production. Thus, while there may be a dominant mode of organisation in a certain region, we can expect to find some farms which are more representative of the earlier dominant systems. Right away, this should caution us not to expect farmers to react in the same way to a particular set of stimulii.

It is not therefore surprising to find a variety of socio-economic structures of production, especially around the long-established European cities. Around Paris, for instance, the large, open plateau and plain-like features have become the domain of large-scale, capital-intensive arable farms (Bryant, 1984a; Phlipponneau, 1956). While still organised dominantly on the basis of individual families, this is one of the areas of France where the development of a more capitalistic form of agricultural production has progressed the most. At the same time, there are areas of more intensive and smaller-scale production of vegetables and fruit, especially in the valley areas north, west and south of the agglomeration, which have their origins partly in a peasant agriculture developed to serve the urban market for fresh produce (Préfecture de la Région d'Ile-de-France, 1988). This artisanal form of production, also called petty commodity production, represents the family-based farm system, with production both for the family's own consumption and for the local and regional market, and with the family providing the labour and capital required for production. Even here, however, a small number of capitalistic farms can be found.

Finally, the systems perspective provides one explanation of why we can find vestiges of former systems present on 'modern, capitalistic' farms, such as some production for family consumption. The importance of recognising the existence of different socio-economic modes of farm production has been reiterated in many studies of agriculture. Recently, for instance, Pomeroy (1986), in her study of structural change in New Zealand pastoral farming, distinguished between capitalist family farmers and simple commodity producers mainly in terms of farm structure, labour processes and farm management behaviour.

What does all this suggest for agriculture in the City's Countryside in the post-industrial era? First, two words of caution. On the one hand, of course, we must underline the point that the threefold categorisation of systems outlined above is a gross simplification. In fact, what we have is a series of modified 'pure' systems representing different degrees of modification, which are conditioned by specific sets of regional, local and individual circumstances, rather than the rather monolithic 'pure' systems that we used as the vehicle for our discussion. On the other hand, we cannot assume that the key forces are necessarily associated in the post-industrial age with the urban-industrial complex; contact with changing values is no longer a simple question of proximity. We shall argue, however, that some aspects of the changing socio-economic organisation are more developed in the City's Countryside because of the greater intensity there of some of the driving-forces in post-industrial society.

What is important is that as the broad structure of society changes, so does agriculture as part of that society. The conditions in which agriculture functions change, and this may elicit a response from farmers (both actual and potential – e.g. the sons and daughters of existing farmers and new potential entrants to the farming sector), either to adapt to and/or react to the changing conditions or to innovate, i.e. to move in a different direction. Some conditions may change throughout the whole geographic system, whether it be the City's Countryside or not, while other conditions are more intense or more fully developed in the City's Countryside.

The forces associated with post-industrial society can now be linked with agriculture and agricultural change in a more explicit way (Figure 1.4). An attempt has been made to separate those forces and responses which are more reasonably seen as the continued development of patterns initiated in the industrialisation of society from those which appear to be more characteristic of post-industrial society. Remember again that while we have separated different types of responses according to a geographic scale of analysis, the

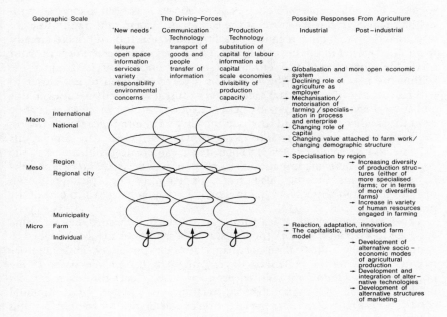

Figure 1.4 Post-industrial society and agriculture

macro-scale responses are built up essentially from the collective responses on the micro scale.

A number of responses from the agricultural sector which continue to be observed today have their roots in the processes of industrial society. In particular, urbanisation and industrialisation, associated with increased job opportunities and better life-style opportunities in the urban-industrial complex and the substitution of capital in the form of more mechanised and motorised techniques for labour input in agricultural production, have given us the capitalistic, industrialised model of agricultural production. This model of agricultural production is characterised by larger scales of agricultural production, more specialised farming both in terms of process (technology) and enterprise structure, and an increasingly important role for capital reflected in equipment and installations on the individual farm unit, as well as strong links between the agricultural production sector and the agri-business supply and processing sectors (Lowe et al., 1990; Troughton, 1982a). Coupled with improved transportation technology and the development of a more open global system for the production and trade of agricultural commodities, we can also see the tendencies noted by various authors (e.g. Chisholm, 1962; Grigg, 1984; Wong, 1983) towards greater regional specialisation in enterprise type as resulting from these processes that evolved out of the industrial society of the nineteenth and early twentieth centuries.

Other forces and responses appear more characteristic of post-industrial society. The emergence of new 'needs' in society has been associated with major changes in the structure of the economic system, although these are also linked to changing technologies of production too. We have already mentioned earlier in this chapter how some of these new needs are at the root of the development of the settlement structure in the City's Countryside within the regional city. Needs such as a perceived need to get closer to nature, to escape the urban 'rat race' – even if only temporarily in the evenings and at weekends – have led many people to choose to live in the countryside or in the smaller villages and towns in the City's Countryside, thus bringing the agricultural populations into closer contact with non-farm populations (Mainié and de Maillard, 1983). Their needs can create problems, e.g. the concerns regarding incompatibilities and conflicts between farm and non-farm populations, but they also create opportunities, e.g. 'pick-your-own operations' in which access to 'nature' or at least open space is also part of the product being sold.

The same needs mean, at least for a part of the population, that the value attached to agricultural work increases, unlike the values that have helped to contribute to the industrial model of agricultural production. This therefore has given rise to the enhanced development of certain elements in the population engaged in agricultural production, e.g. part-time farming and hobby farming. Changing values relating to greater environmental consciousness have also led to the adoption of conservation-oriented farming practices on some farms.

In many cases, the intensity of the forces, either in terms of markets for agricultural products and services or in terms of the supply of potential new types of entrants into farming, appear greater in the City's Countryside than elsewhere. Consequently, we should not be surprised to see in some parts of the City's Countryside an increasing variety of socio-economic modes of production and an increasing variety of production or enterprise structures, either in terms of a greater diversity of specialised farming or of the enterprises present on a given farm. What is interesting from the perspective of the evolution of agricultural systems in the City's Countryside is that capitalistic farming with an industrial orientation is not precluded from participating in the opportunities created by some of the new 'needs' of post-industrial society. A capitalistic mode of decision-making therefore does not have to go hand in hand with an industrialised mode of production; indeed, in some respects capitalistic operators may be more inclined to take advantage of such evolving opportunities.

Information as a need and as an increasingly important component of agricultural production shows up in the greater levels of management abilities now required in farming. In the context of conservation technologies and practices in farming, better understanding of the links between the biophysical resource base and the impacts of farming practices has become increasingly recognised as critical to the long-term sustainability of our food production system. Low input farming techniques, minimum and zero tillage practices, organic farming and alternative agriculture are among the various terms to which we have been exposed with greater frequency in the last part of the twentieth century. This is a significant shift, even if it is not yet widely accepted and integrated into farming systems which still function on the basis of the industrial model. For those farmers who adopt such technologies, while there may be an immediate economic incentive that some perceive in this, the driving-factor in such developments is more likely to reflect a value shift for those operators – a more acute awareness of their link with the biophysical environment and the development of a particular conservation ethic towards the land resource. All this, as with the changing values associated with the development of new sets of needs generally, means that there appears to be a growing acceptance of a variety of alternative production systems and socio-economic modes of production with their different life-styles (Chassagne, 1980).

Some of the changing conditions still facing agriculture simply represent a continuation, and sometimes a development, of patterns that were established at the height of industrial society, and therefore some conditions continue to support continuing modes of production. However, our discussion suggests that the capitalistic, industrialised model of agricultural production is just one particular path along which agriculture has evolved, and can continue to evolve. It is unquestionably an important component of our agricultural production system, and its development has been encouraged and supported in many ways by public institutions (government, agricultural research colleges, educational programmes) as well as agribusiness (Dahlberg, 1986). However, it is only one of the possible paths; there have always been a variety of modes of organisation of agricultural production, as our earlier discussion implies – the variety has simply not always been recognised. Furthermore, with the pressures and transformations that have accompanied post-industrial society, other modes of organisation have evolved: variants of part-time farming, hobby farming, organic and alternative agriculture.

It is important to recognise these different modes of organisation of agricultural production because they imply that farmers will not necessarily respond in the same way to the same set of stimulii. This is not just because of individual idiosyncracies, although they exist too, but because different modes of organisation are associated with different sets of motivations or driving-forces. Furthermore, farmers do not all belong to the same system of exchange. For instance, in New Zealand, Canada and elsewhere, a variety of producer marketing boards exist, each involved in a different system of exchange which have influenced different producer groups to evolve along somewhat different lines. Finally, variation can be expected geographically in responses from a given mode of organisation of agricultural production to the extent that the level of intensity of the stimulii, pressures and opportunities vary geographically. This is what lies behind Bryant's (1984a) suggestion that the dynamics of agricultural landscape (and structural) change in the City's Countryside could be represented schematically as a threefold categorisation into landscapes of agricultural degeneration, landscapes of agricultural adaptation and landscapes of agricultural development; we shall return to this scheme again in Chapter 4.

Variation therefore comes from several sources: different modes of organisation of agricultural production: participation in different systems of exchange; differential patterns of stimuli, pressures and opportunities, different operating environments; and individual or personal factors associated with the farmer and farm family. It is not surprising that we observe complex patterns of responses and agricultural structures in the City's Countryside. What emerges from this discussion is an agricultural mosaic of overlapping structures, modes of organisation and responses, which underlines the complexity of agriculture in the City's Countryside. It is essential to recognise this complexity and its underlying causes if we are to understand agriculture in the City's Countryside, and to stand any real chance of integrating the values of the various individuals and families who are engaged in agricultural production with some of the broader collective values associated with agriculture and agricultural areas in the City's Countryside. Rather than frowning on this complexity, we can see it in a very positive light for, as Mainié and de Maillard (1983) pointed out, the areas near cities are veritable social laboratories; all sorts of life-styles and socio-economic modes of production which are tried out and coexist there will eventually shape our futures.

Functions of and interests in the agricultural land resource

We have already stressed the variety of interests in agricultural land even from the perspective of those engaged in agricultural production activities. For some it represents primarily a means by which they are able to derive their livelihood; for others it provides a necessary supplementary source of income; for others it represents the means for satisfying personal needs for working close to the natural environment; and for still others it provides the basis for a leisure-time activity. This is complicated by the different perspectives with which individuals view the land resource: as simply an input into the production process (and here, either short or long-term perspectives can be taken) or as a resource that requires nurturing to maintain its productive potential.

The agricultural land resource also has other values associated with it, partly because of the variety of functions that can be linked with agricultural activity. Some of the more common functions are listed in Table 1.1, and they are categorised loosely according to whether they are dominated by the production, protection, place or play function of land (Bryant et al., 1982; Russwurm, 1977). First, of course, the land resource has a value because of what it can PRODUCE for sale in the market-place. Agricultural produce, both food and non-food produce, and services that can be linked fairly closely with an agricultural activity, e.g. livery stable, all involve the exchange of produce or services for money. A monetary value is therefore attached to the land resource involved. We could also extend this to include the addition of non-agricultural activities or enterprises on the farm which utilise under or unused farm resources, e.g. use of obsolete barn space for parking recreational vehicles.

Table 1.1 Functions of agricultural land: examples

Broad functions	Private interests	Collective interests
PRODUCTION	Farm level production – food – non-food products – services	Potential food production Access to food supply Support of food processing and agricultural supply industries
PROTECTION	Private reserves	Wildlife habitat support Water supply
PLACE	Housing Industrial and commercial development Agricultural production oriented to specific markets	Access to housing Support for economic development and employment Infrastructure support
PLAY	Recreational enterprises Hobby farming Recreational enterprises on farms	Amenity value of agricultural landscapes

There are also broader collective values associated with the production function of the land. They include the potential of the land resource when incorporated into a functioning agricultural system to contribute to feeding the population (nationally, internationally), and the role the agricultural production system plays in supporting other economic activities, such as the food processing and agricultural supply industries (Bryant and Russwurm, 1979).

PROTECTION functions and values of the agricultural land resource are primarily associated with collective values linked to other elements of the environment upon which agricultural activities may have an effect. Most notable here are the potential impacts that agricultural activity can have upon wildlife habitat and water supplies (both quality and quantity) and valued aspects of the cultural environment or those amenity values associated with agricultural land because of the cultural and historic values that are often embedded in the agricultural landscape. All of these protection functions have long been explicit concerns in much of Western Europe (e.g. Blunden and Curry, 1988). In the Netherlands, for example, not only does the agricultural landscape often contain important links with history and culture, but the nature of the ecosystem is such in some areas that agricultural change can have a dramatic and fairly immediate impact on water quality and wildlife (cf. van Oort, 1984a).

PLACE functions concern the value that land has because of its location. They mainly involve non-agricultural functions, such as housing, industrial and commercial development, although there are certain agricultural activities in the City's Countryside for which location is at least as important as the inherent productive capability of the biophysical environment. Where the interest is expressed through the acquisition of a proprietary interest in the land, the land acquires value through the land market. However, there are also collective values or interests in the land for these functions since they all represent potentially legitimate needs which have to be met somewhere. Therefore, agricultural land in certain parts of the City's Countryside has the potential to support some of these functions, and any planning or management must take these potential functions of the land into account.

Finally, PLAY functions relate to recreational or leisure-time activities and pursuits that are supported by agricultural land or agriculture. On the one hand, some agricultural activities are principally undertaken as a hobby and are therefore recreational and leisure-oriented. Other recreational activities may be integrated with or tacked onto a functioning farm unit as supplementary enterprises, e.g. campsites, livery and riding-stables. At the collective level, where again we are dealing with non-market oriented functions, agricultural land and agricultural activities support a landscape in which are embedded various elements that may be valued by society or at least by certain parts of it: aesthetically pleasing landscapes, wildlife habitats, and elements and sites of historic and cultural value (Coppack et al., 1988). So, even where agricultural production is the dominant function, many agricultural areas also support other functions, including the general one of providing a 'breathing space' for the city (Crabb, 1984).

In view of all these values, functions and interests in the land, many of which we can expect to be much more intense in the City's Countryside, the 'foodland' perspective in which the agricultural production of foodstuffs is seen as the most important priority in the management and planning of the agricultural land resource is rather a narrow one. Where the issues of agricultural overproduction have been most acute, e.g. in the European Community, the agricultural production for food perspective has been in rapid retreat during the 1980s, and the other functions of agricultural land have attracted more attention in the ongoing policy debates (Blunden and Curry, 1988).

Management and planning for agriculture in the City's Countryside

Management and planning for agriculture in the City's Countryside has been dominated by considerations regarding the land resource. The variety of structures and of values, interests and functions associated with the agricultural land resource raises the questions: How should we manage agricultural land? Which values do we take into account? For whom do we do it? The 'foodland' perspective – a sort of agricultural fundamentalism (Blunden and Curry, 1988) – provides a simple answer, because it leads to an absolute priority being assigned to conserving the agricultural land resource when taken to its logical conclusion. But the world is not that simple!

The challenge that exists for managing and planning agricultural land in the City's Countryside is to be able to render compatible the protection and development of the collective values or interests in the land resource with the values of the people who work the land for their living. This means that management and planning must also take into account the people and families involved, and the business and other organisational considerations involved in functioning farm operations.

The key collective values deal with

1. agricultural land as a renewable resource and the contribution that it can make to feeding present and future generations;
2. agricultural land and activities as a supporter of amenity values in the landscape; and
3. agricultural land and activities as a support of and link with natural environment elements such as wildlife and water quality.

Our attention has been increasingly drawn to the notion of 'sustainable development', a construct that can be usefully applied to agricultural activities and the necessary resource base in all three of these collective values. Focusing on the food production aspect only for the moment, Brklacich (1989) has shown how even striving towards a sustainable food production system requires consideration of dimensions besides the environmental quality of the biophysical resource base for agricultural production. These other dimensions include the sufficiency of the system to provide for the needs of the population in terms of foodstuffs (clearly, a dimension that links directly with the agricultural land conservation movements and programmes in many parts of the Western world), as well as the ability of the production system to return a satisfactory level of return and livelihood to the farm producers involved (Brklacich et al., 1989; Brklacich et al., 1991).

All the issues relating to the collective values of the agricultural land resource and agricultural activities can be placed into the sustainable development perspective in the City's Countryside: farmland loss, indirect impacts of urban development upon agricultural viability, soil and water degradation, and landscape amenity. The collective interests in the land resource involve a number of difficult questions since those for whom the land is valuable are not just the owners or occupiers of the land and in many cases are not even resident locally or even regionally. One of the largest continuing challenges therefore is to integrate these broader values and interests into the systems of planning and management for the City's Countryside, which are generally more locally or regionally-oriented. Who benefits from public sector intervention in the management and planning of agricultural land and activities and who pays for it? Answers to such questions become more and more complicated as soon as we recognise that there are few absolutes: different weights and values are attached to different interests and functions of agricultural land and activities, and these vary at different times. In the following chapters we shall try to extract some of the key relationships and present a view of planning and management of agricultural areas in the City's Countryside which advocates the incorporation of all of the legitimate interests in the *process* of managing change.

2
The resource base: the underpinnings

The agricultural land resource in the City's Countryside

Since prehistoric times, agriculture and cities have been inextricably interlinked (Benevolo, 1980). Not only was agriculture a cornerstone upon which urbanisation first emerged (Bairoch, 1988), but throughout history, agriculture has contributed significantly to the world's economic and urban development. The demands for labour during Britain's Industrial Revolution were, for example, met to a large extent through the technological developments in agriculture that released large numbers of people from agricultural labour. In Canada, Quebec's low agricultural productivity has been held responsible to a large degree for the slow rate of growth of the province's urban-industrial complex compared with that of its neighbour, Ontario (McCallum, 1980). The rate of change in agriculture has therefore been linked to changes in the urban-industrial complex. The relationship is, of course, a two-way one, with urban-industrial development drawing labour away from primary production as well as changes in agricultural productivity releasing labour.

Another expression of the special relationship that has existed and continues to exist between agricultural and urban areas is seen in the fact that many of the world's cities are surrounded by good quality farmland (Hill, 1986; OECD, 1976). This generalised land use pattern has been observed in many countries, including the United States (Hart, 1976; Vining et al., 1977; Zeimetz et al., 1976), France (Bryant, 1981), Great Britain (Best, 1978; Cruickshank, 1982; Embelton and Coppock, 1968; Wibberly, 1959), and New Zealand (Land Use Advisory Council, 1983; Meister, 1982).

In Canada, a country long thought of as having an abundance of land, the geographic coincidence of urbanised areas and good quality farmland has been emphasised many times (Manning and McCuaig, 1977). Neimanis (1979) has estimated, for example, that 57 per cent of the nation's Class 1 agricultural land (based on the Canada Land Inventory system) is located within an 80 km (50 mile) radius of its twenty-three largest cities. Even more striking is the fact that almost 25 per cent of Canada's 2.4 million hectares of Class 1 farmland can be found within 80 km (50 miles) of just one city – Kitchener, Ontario. And, as Simpson-Lewis et al. (1979) point out, a person standing on the observation deck of Toronto's CN Tower on a clear day can 'see' 37 per cent of the nation's Class 1 land for agriculture! The general relationship in Canada is therefore quite strong. The closer to an urban centre in Canada, on average the higher the proportion of good quality farmland (Neimanis, 1979). Within a 40 km (25 mile) radius of Canada's largest cities, 51 per cent of the farmland area is prime for farming; but within a 160 km (100 mile) radius, less than one-third of the farmland area is classed as prime (Table 2.1).

Australia is another country where the population is concentrated in areas that are well suited for farming. Australia's population is densest in those regions where farmland yields the highest economic return per hectare (Figure 2.1).

Table 2.1 Agricultural capability of land around Canada's 23 largest cities

Radius in km	Percentage of land prime for agriculture
40	51
80	43
121	36
160	32

Source: Calculated from Neimanis (1979, Table 1)

Urban expansion and farmland conversion

Given the fact that many cities are surrounded by productive farmland, it is not surprising that when cities expand, they frequently take over good quality farmland. This has prompted widespread concern over the 'permanent' loss of a renewable resource. Some have expressed the fear that the urbanisation of farmland may result in future food shortages (Agricultural Institute of Canada, 1975; Ontario Institute of Agrologists, 1975), which would in turn be detrimental for balance of trade figures (Bryant and Russwurm, 1979), or lead to increasing food imports from sources which may be unstable or unreliable in terms of price and/or supply. Still others have expressed concern over the urbanisation of farmland because of the amenity value many farmscapes possess (Krueger, 1977b).

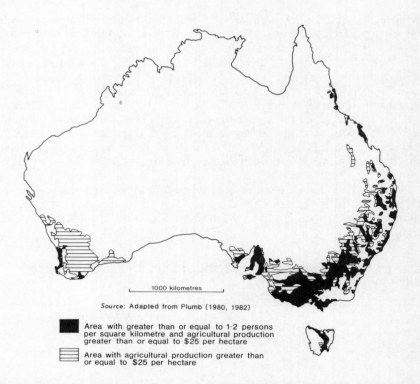

Figure 2.1 The spatial coincidence of population and highly productive agricultural land in Australia, 1976

Consternation over the loss of farmland to urban use can be traced back to the 1930s (Best, 1981). The ensuing debate has attracted the attention of planners, politicians, scholars and others. It has found its way into both academic journals and the popular media, and in the 1970s in North America, when concern was perhaps most intense, the issues were hotly debated.

The debate cooled somewhat during the 1980s, largely as a result of the recession suffered by many countries. However, as many of the larger Western cities appear to be experiencing a period of renewed expansion, we may well find ourselves immersed in yet another round of debate. While the debate concerning the loss of productive agricultural land seemed to wane in the United Kingdom by the 1970s, renewed fuel for the debate comes from concern about the real costs of maintaining the high levels of agricultural productivity through intensive farming practices.

Finally, on the continent of Western Europe, while concern over the loss of the productive land resource has rarely seemed to attract much attention, costs of current levels of intensification (e.g. Kohn, 1990) and the amenity functions of agricultural landscapes are such that anxiety over the impact of urban development upon agriculture is likely to increase. Before we delve into the various perspectives introduced above, some commentary on the magnitude of the conversion of agricultural land to urban uses is required.

Evaluating the reliability of land conversion estimates

How much agricultural land is involved in land conversion? The amounts of land which have shifted out of agriculture and into urban uses have been enormous in some countries. The country which has experienced the most farmland conversion in absolute terms is undoubtedly the United States, where between 1910 and 1959 an estimated 7.5 million hectares of agricultural land were transferred to urban uses (see Table 2.2). In England, where much apprehension stemmed from rapid urban expansion in the period following World War II and in the context of the vulnerability of the food supply in times of international stress, Best (1981) reported the conversion of almost 1 million hectares of farmland to urban use over a period of nearly sixty years ending in 1979. By comparison, the amounts of farmland converted in smaller countries such as Scotland, Canada and New Zealand appear small, but the urbanisation of farmland in relation to their total land base has been an issue of debate in each of these countries nevertheless.

Table 2.2 Examples of estimates of the conversion of agricultural land to urban uses in selected countries

Period	Country	Area Converted '000 ha		Reference
		Total	p.a.	
1922–79	England & Wales	887.5	15.3	Best (1981)
1910–59	USA	7,500.0 (farmland)	150.0	Hart (1968)
1966–86	Canada	301.4 (total rural land)	14.4	Environment Canada (1989) (centres over 25,000)
		174.8 (prime farmland)	8.7	
1971–79	Scotland	10.0	1.1	Department of Agriculture and Fisheries for Scotland (1982)
1949–66	New Zealand	16.2	0.9	Molloy (1980)

Best and Coppock (1962) pointed out that whether or not one believes that agricultural land lost to urban development is a problem, it is essential that the basic facts which feed the debate be sound. This has not always been the case, partly because the issue has become so emotional, and partly because of certain biases both in conceptualising and measuring the phenomenon.

The first problem, undoubtedly the most serious to cloud the debate, is that many researchers do not describe in detail the methodology used to derive their figures. Moreover, it has been pointed out that many statements concerning the amount of agricultural land taken over by the advancing urban margin are 'strictly anecdotal', while others seem to be 'little more than sheer guesswork' (Hart, 1976, 5–6). Much of the work, that is difficult to substantiate in any objective and systematic way has appeared in various in-house reports, political party pieces and newspaper reports. The real problem arises when students, other researchers, policy-makers and journalists treat all published material with equal credibility, irrespective of how the evidence was established.

A second problem involves the frequent assumption that all adjustments in farmland area in the City's Countryside are the result of urban expansion (see, for example, Bogue, 1956; Crerar, 1963). When this assumption has been made, and where the measurement technique used is an indirect one (see below), the amount of farmland urbanised is overestimated. The issue is essentially one of conceptualising the phenomenon of the removal of land from agricultural production, and understanding the limitations of different measurement techniques and data sources.

Conceptually, agricultural land in the City's Countryside can follow several paths (Figure 2.2). Clearly, farmland can be withdrawn directly from normal agricultural production and put immediately into urban uses, or it can pass through various stages of disinvestment while still supporting agricultural use before being converted. Farmland may even be set aside before being finally converted to urban uses. (The topic of disinvestment in farmland and the other changes that can be induced in agricultural land (extensification, intensification) by different combinations of urban and non-urban forces are treated in detail in Chapter 3).

The term set-aside ('idle land' is used in North America) refers to land that is temporarily taken out of agricultural production during the transition of land from one economically and/or socially productive use (agriculture) to another (e.g. residential development). It is possible that conditions surrounding the transition change so that the set-aside land could revert to agricultural use or could be finally abandoned, a state which exists when the land essentially has no foreseeable economically and/or socially productive use. Land can also be withdrawn from agricultural production due to shifting economic margins as it becomes uncompetitive; abandonment of the land then occurs, although it is possible for the process to be reversed again. Furthermore, some land withdrawn from immediate production to be put to certain other uses is not necessarily lost to future agricultural production, e.g. parkland and golf courses, and this is an important point when we focus attention on the potential of the production system rather than on actual

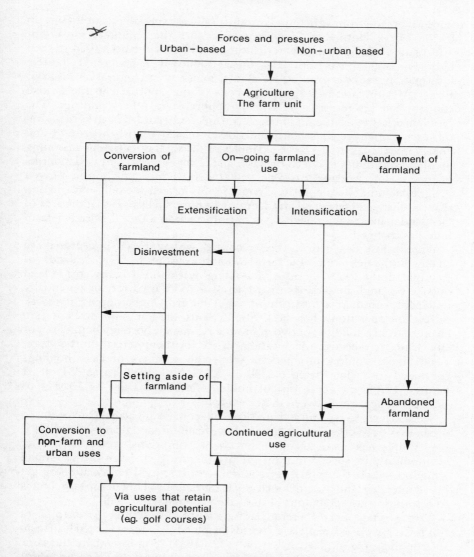

Figure 2.2 The removal of agricultural land from agricultural production

volume of production. The phenomenon of the removal of land from agricultural production is not therefore a simple process, and in measuring the conversion of agricultural land to urban or non-farm uses, it is important to ensure that the technique and data source used do not confuse land conversion, set-aside and abandonment of land.

The interpretative problem arising from the confusion of different types of removal of land from agricultural production was especially troublesome up to the 1970s. Until then, and frequently since then, indirect sources of information on the changing agricultural land base have been used to shed light on land conversion. The most frequently encountered indirect information source is from official statistics such as agricultural census data. Best (1981) discussed the pros and cons of this approach in the British context, but his remarks are equally applicable to other countries. The problem arises when changes in the farmland area recorded in an agricultural census are interpreted as farmland losses related to urban growth and development. The now classic examples of the sorts of fallacious reasoning involved are studies in which the farmland changes were expressed in terms of the area of farmland losses per 1,000 increase in population in a given geographic unit (Bogue, 1956; Crerar, 1963). In fact, the only information that the change in farmland area from such sources tells us is that the area of farmland managed within whatever the information source defines as a 'farm unit' has changed!

Where a 'loss' or a 'gain' has been recorded, we have first to be certain that a change in the definition of a 'farm' has not occurred. In New Zealand, for example, in the 1969–70 census, a 'holding' was defined as any area of land outside borough boundaries of ten acres or more used for, or potentially usable for, commercial horticulture, vegetable growing, cropping, livestock or forestry operations. For the 1970–71 census, the minimum size criterion changed to any 'holding' of two or more acres outside borough boundaries or any land, irrespective of location, used for commercial horticulture, vegetable or poultry production. While the numbers of farm holdings continued to decline (from 66,380 on 31 January 1969, to 65,331 on 31 January 1970), the total area of holdings increased from 17,350,174 to 17,432,110 hectares between these two dates.

We also have to recognise that a 'loss' can come from actual conversion to urban uses or to other uses (e.g. forestry), from the land being set aside or it may simply have been abandoned without being transferred to any other economically or socially productive use (Figure 2.2). Even in the United Kingdom where the annual agricultural returns request information on the broad intended land uses to which any land transferred out of the farm unit is to be put, misinterpretations of these data abound (Best, 1981).

Census-type data are therefore far too easy to misinterpret. An example of this is Crerar's (1963) analysis of census farmland change in selected Canadian metropolitan regions; when Bryant et al. (1981) took essentially the same regions but disaggregated the regions for the analysis of the farmland census data, enormous internal variations in the apparent rate of 'loss' of farmland per 1,000 increase in population were uncovered, a variation that could not be explained by any real variation in land conversion.

Even in some of the better agricultural regions, there are often areas of poorer agricultural land, and economic and technological change in agricultural competitivity can take their toll in such regions as well as in the more marginal agricultural areas. When the whole range of biophysical environments that exist for agricultural development around cities in many countries is taken into account, it is quite clear that forces pertaining to changing technological and economic competitiveness have played an important role in farmland removal. Bryant and Greaves (1978) show, for instance, that the relative rates of census farmland removal around Canadian metropolitan centres from 1961 to 1971 were actually greatest around those centres where urban population growth was smallest, i.e. around the Maritime cities and the northern Ontario and northern Quebec cities. These regions coincided with some of the poorest agricultural land in the country.

More direct methods of measuring the rate of agricultural land conversion such as through field work and air photo analysis have been used for a long time by geographers in the analysis of small areas. Large-scale exercises based on mapping programmes such as the Land Utilisation Surveys in the United Kingdom have always met with difficulties due to timeliness and stability of classifications over time. Even careful air photo analysis in a single region can encounter problems when the change analysis is undertaken at two different times by two different teams of interpreters. In the Ile-de-France region, for instance, based on two independent air photo analyses of land use in the region conducted for 1974 and 1982, the annual 'consumption' of agricultural land was estimated at 4,400 ha (IAURIF, 1987). The authors of the report hasten to add that this includes abandoned lands, and furthermore, readers of the initial report are warned about the increased detail utilised in the 1982 analysis and of the difficulties of direct comparison with other sources of information on land conversion because of definitional differences (IAURIF, 1987).

Larger-scale analyses of a national scope using air photo analysis have undoubtedly advanced our knowledge of the volumes of actual land conversion and the relative importance of different types of land use change. A US study of land use dynamics in fast growth urbanising counties documented considerable regional variations in the link between urban expansion and agricultural land adjustment (Ziemetz et al., 1976). Some urbanising counties such as Florida actually experienced net increases in cropland, and in others, afforestation and abandonment were identified as important components of farmland removal. Changes in relative agricultural productivities between regions linked to the differential impacts of technological change in agriculture were identified as a major factor in the cropland decline in the urbanising counties studied.

National-scale investigations of land conversion in the City's Countryside using air photos do face problems however, particularly when there is a lack of a common temporal coverage of air photos. In some Canadian studies, this necessitated extrapolating data backwards and forwards to common points in time to achieve 'comparability' between different regions (Gierman, 1977; Warren and Rump, 1981). This can pose additional difficulties in some regions if the rates of development vary substantially over relatively short periods.

Notwithstanding such cautions, the direct approach to measuring agricultural land conversion rates is much more satisfying and has contributed enormously to correcting previously inaccurate images, especially in North America (see especially the direct approaches used in Gierman (1977) and Warren and Rump (1981) for Canada and in Ziemetz et al. (1976) for the United States). Thus, Bryant et al.'s (1981) comparison of Gierman's (1977) study, which was based on direct measurement with their own data based on indirect measurement (census data), highlighted the problems associated with the earlier attempts to measure farmland 'loss' using census data. Gierman (1977) estimated from air photo analysis that in Canada around cities of over 25,000 in population between 1966 and 1971, 69.6 ha of rural land were converted to urban uses for every 1,000 increase in population in those urban regions. In contrast, for the same period using census data for assessing farmland change within a 50 km (31 mile) radius of urban centres with a 1976 population of 40,000, the rates of change were substantially higher than Gierman's (2,447 ha per 1,000 increase in population around the Maritime provinces' urban areas, 1,201 ha for Quebec and 522 ha for Ontario).

Even though there are methodological differences between the two analyses (a slightly different population of urban centres and a broader 'fringe' zone for the census data analysis), the high values revealed through census data cannot be explained even to a large degree by land conversion. Nevertheless, despite such advances in our knowledge, census data still continue to be used and misinterpreted. An example is Allison's (1984) analysis of the Golden Horseshoe counties in southern Ontario, in which urban growth forces were interpreted as practically the sole cause for the 30 to 47 per cent loss in their census farmland base from 1941 to 1976, despite the fact that there exists in this area some significant areas of technologically and economically marginal farmland upon which farmland abandonment has occurred.

The availability of satellite imagery essentially since the mid-1970s has added new possibilities to the quest to measure land conversion in the City's Countryside (Bryant et al., 1989b). The use of official statistics, especially with agricultural census and the like, is usually dogged by the representation of only indirect evidence of land use change, by its relatively infrequent availability and by the lack of availability of detail in the case of most national censii for up to two to three years after the census is taken. Air photo analysis turns out to be costly in terms of human resources as well as acquisition of the data. Furthermore, the data are not usually available simultaneously for all urban regions in a country – countries with small national territories obviously represent an exception. In view of the limitations of official statistics and air photos, the use of satellite imagery presents some considerable potential advantages: frequent and timely availability, broad spatial coverage and relatively low cost (Bryant et al., 1989b).

However, there are some limitations, especially in the use of available imagery during the 1970s and early 1980s (Fung and Zhang, 1989). The limitations deal principally with the degree of spatial resolution of the earlier imagery and generally the ability to classify land uses with a reasonable degree of accuracy. The more recent generation of satellites (e.g. the Landsat Thematic Mapper and the French satellite SPOT) have effectively resolved the spatial resolution problem with a much finer geographic detail (Johnson and Howarth, 1989), and there is a much greater degree of detail provided for each geographic element or pixel. However, the degree of accuracy of land use classifications, especially for multipurpose classifications of the type necessary to evaluate the multiple land uses and changes in the City's Countryside, remains problematic.

Advances continue to be made in the resolution of such issues, such as the use of ancillary information to improve the accuracy of estimates of change from satellite imagery through stratification and the use of extensive patterns of control points. Further improvement of the integration of ancillary data and remote sensing data through the development of powerful Geographic Information Systems hold out even more hope for the monitoring and evaluation of land use changes on a regular and frequent basis. However, these types of advances so far have generally been restricted to the study of relatively small areas, e.g. a given urban fringe area, and ways of circumventing the high costs of developing extensive systems of ground control for multipurpose projects will have to be developed or alternatives found.

A study by Environment Canada (1989), updating their earlier work on monitoring land use around Canada's urban centres (Gierman, 1977; Warren and Rump, 1981) has demonstrated the utility of working with several data sources. While the land activity/land cover information for 1981 was developed using air photos combined with some field checking, the extent of the urban boundary was developed mainly from using Landsat Thematic Mapper transparencies. This demonstrates that satellite imagery can still be used manually in interpretation, and when the purpose is focused (i.e. in this case, identifying the extent of the urban boundary), the results can be accepted with a high degree of confidence.

Another point must be made about the identification and estimation of the areas of land converted to urban and urban-related uses, and this relates primarily to direct approaches to measuring the conversion through air photos and satellite imagery. There are different processes of conversion of agricultural land to urban and urban-related uses. It is useful to distinguish two basic processes and forms, first, accretionary development at the edge of cities, and second, the more scattered type of development. Accretionary development is fairly easy to identify at the edge of cities, particularly using a combination of air photos and satellite imagery. Residential, commercial and industrial development can be identified, and so can urban-related uses such as parks, hospital grounds, golf courses and other uses that include large areas of green open space because of their proximity to the more traditional urban uses. Even so, careful ground truthing of satellite imagery analysis through field checking and/or use of air photographs and other ancillary information is important in achieving acceptable levels of accuracy. On the other hand, the more scattered types of urban and urban-related development, being spread over a much larger area, are much more difficult to identify in the context of national analyses because the potential for confusion between uses is that much greater and the inability to rely upon proximity to other land uses to pin down the classification is a distinct drawback.

Finally, this discussion regarding different types of conversion raises the issue of development densities and the factors that account for observable rates of land conversion. Population growth in specific metropolitan regions is certainly a key element in accounting for some of the land converted. In addition, however, redistribution within a given metropolitan region (such as in the Paris region, France) and the relatively low densities of development, especially in North America, are at least as important in understanding the overall magnitudes of conversion (Best, 1981).

An underlying concern with the efficiency of development is one of the reasons why many studies have sought to express land removal from agriculture (indirect approaches) or land conversion (direct approaches) in terms of amount of land involved per unit increase in population (e.g. Bogue, 1956; Environment Canada, 1989; Gierman, 1977). The studies cited have demonstrated significant regional variations in the rates of land conversion, some of which are related to certain urban regions 'catching up' with others in terms of infrastructural provision, such as in some of Canada's Maritime urban regions (Gierman, 1977). Another major component is simply the efficiency with which land is used. The sequence of studies undertaken by Environment Canada on monitoring land use around 70 urban regions with populations over 25,000 shows that the larger the city, the more efficient the land conversion, a relationship that remained constant from 1966 to 1986 (Environment Canada, 1989). Thus, for the cities with populations over 500,000 during this period, only 50 ha of rural land were converted for every 1,000 increase in population, compared to 78 ha for cities between 250,000 and 500,000, 101 ha for cities between 100,000 and 250,000, 175 ha for those between 50,000 and 100,000 and 196 ha for those between 25,000 and 50,000. Clearly, in appreciating the overall rates of change, it is imperative to investigate the conditions under which land is converted (development densities, development standards including park space provision, and the like), and not just focus on population change.

Assessing the significance of the urbanisation of farmland

Advances have been made in measuring the conversion of agricultural land to urban uses; even so, some vexing questions remain. Knowing the amount of land converted does not address the issue of its significance and whether it warrants some form of intervention. Farmland is, of course, fundamental to most agricultural production. It is certainly partly because of this that any conversion to urban uses has given rise to concern in some quarters over its impact on the ability of the agricultural food production system to respond to food needs, a critical dimension of sustainability of the food production system. However, the issue is much more complex than simply calculating the lost production due to the actual conversion of land to urban uses. We propose to discuss the different dimensions of this problem below in order to emphasise the linkages between agriculture in the City's Countryside and other parts of the socio-economic system and to set the scene for the analysis of public intervention in agricultural land use management and planning in Chapter 5.

Perspectives taken on the loss of farmland to urban land uses range from outright condemnation of practically any loss of farmland (the 'every acre counts' perspective) to unwaivering faith in the market system in allocating scarce resources (in this case, the agricultural land resource) to alternative uses in an efficient manner. These represent the two extremes, and both are naïve. A number of themes exist in the debates over the loss of farmland to urban uses (Table 2.3). Our immediate concern is with those involving current versus future (and therefore, potential) food supply. These two food-related themes are present to a greater or lesser degree in most debates over the loss of farmland to urban uses, and both deal primarily with the food sufficiency dimension of sustainable food production systems. (The other two dimensions of sustainable food production systems which were introduced in Chapter 1 are maintenance of the productivity of the biophysical resource base and maintenance or achievement of satisfactory returns to the producers in the agricultural industry (Brklacich, 1989)).

At one extreme, some consider any farmland at all being converted to urban uses to be a significant problem. Underlying such concern is the notion that agriculture is one of the foundations of our society, that it represents everything that is 'good, natural and pure'. Proponents of this view espouse their disdain of urban development with almost fundamentalist zeal. Farming is placed on an unassailable pedestal. The fact that capitalist agriculture has required the simplification of natural landscapes and ecosystems (Dahlberg, 1986), and that many serious environmental problems have been caused by modern agricultural practices (Munton, 1987) seem lost to many of the extremist farmland preservationists.

Others, expressly concerned with current food production, contend that so long as there are food shortages in the Third World, Western nations have a moral obligation to stem the tide of urban development onto good quality farmland (Brown, 1981). This is a difficult argument to question on moral grounds. 'Need' as expressed through starvation is difficult to ignore, but it raises two fundamental concerns.

First, agriculture in the Western world is primarily oriented to satisfying the effective market; this can be expressed through industries and individual consumers purchasing in the 'free' market or through governments and agencies through purchasing programmes for storage or delivery as food aid as well as through influencing the price signals conveyed through market systems of exchange to the agricultural producers. The key point is that supply is tied into the perceived monetary rewards of production.

On the side of those who favour the market mechanism for the allocation of agricultural land between alternative uses, we can point out

1. that over the twentieth century, domestic food supplies have been assured in the West, and changes in agricultural technology have more than compensated for any loss of capacity due to conversion of farmland to urban uses – witness the slow rate of increase in food prices to the consumer relative to other consumer products;
2. that some agricultural land is devoted to the production of non-food items for the domestic market (e.g. horticultural production and turf production); and

3. that there has been a preoccupation in the Western world with managing production capacity that is surplus to effective demand through supply management and quotas for some sectors, government purchase and storage and/or resale schemes.

Table 2.3 Considerations in the evaluation of the significance of farmland conversion to urban uses

Themes	Considerations
1. Current food supply and demand	– The need and market for food – Agricultural surpluses – Supply management – Food prices – Distribution deficiencies – Non-food farm products
2. Future food supply and demand	Supply – Land resource quantity and quality (potential in relation to the total food production system) – Other agricultural resources – Role of food imports, trade balances and dependency relationships – Other forces influencing resource potential and supply capability – Role of technology and intensification – Special agricultural lands Demand – Domestic versus foreign markets – Changes in consumer tastes – Demographic patterns and change
3. Role of farmland in environmental quality	– Link to water quality – Link to wildlife habitat
4. Role of farmland as amenity support	– Link to 'natural' environment – Link to cultural and historical values

On the other hand, we must acknowledge both the artificially low prices of food products because of the cheap food policies pursued by government in many Western countries to benefit the (urban) consumer through such mechanisms as subsidies and deficiency payments, and the long-term negative impacts of agricultural technology on the sustainability of the agricultural land resource (see below).

A second concern is the extent to which the problem of need, or Third World hunger, is really a function of the productive capacity of the food production system *per se*. Other dimensions appear to be at least of equal importance, namely the inadequacy of food distribution systems (Blunden and Curry, 1988; Falcon et al., 1987) and inequity in the access to resources in many Third World countries (Parikh and Rabar, 1981). Faced with those types of issues, farmland conservation policies in industrialised nations will do little to address the issues of world hunger. Even if for a variety of reasons farmland conservation programmes are instituted, the maintenance of agricultural food production capacity requires an effective market for the agricultural produce. This raises the questions of either how the difference between the volume of production which the effective market will absorb at any one time and any higher actual volumes of production will be paid for (and who will pay for it) or how to maintain a production system that has been put 'in reserve' so to speak (and who will pay for that!). These issues go far beyond the scope of land use planning *per se* and can only be resolved through political choices (at all levels) concerning the goals, objectives and priorities of a given country.

While it does nothing to reduce the complexity of the issues that we have begun to raise, the other theme related to food production is focused on the potential of the food production system to meet future market needs in as efficient and effective a way as possible. A key consideration in this theme is the relatively limited supply of good quality farmland overall in a given country and the fact that it tends to be disproportionately represented in the actual volume of conversion of agricultural land to urban uses (Table 2.3).

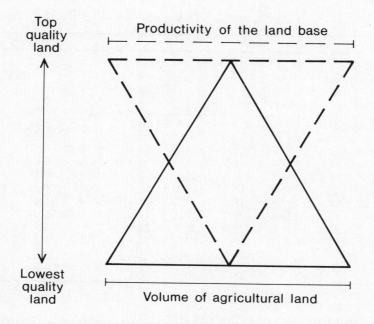

Figure 2.3 Schematic representation of the relationship between the quality and volume of agricultural land

Two generalisations can be made about the agricultural land resource quality consideration. First, there is indeed an inverse relationship between the area of agricultural land and the 'quality' of the land resource (Figure 2.3). The notion of 'quality' is related both to the potential productivity of the land resource and the range of different agricultural enterprises (the 'options') that can be supported economically by the land resource in a given location. The latter dimension can be usefully thought of as representing the potential

Table 2.4 Distribution of agricultural land in Canada by agricultural land capability and by province

	Province% [a] of CLI [b] area by class in each province						
	1	2	3	4	5	6	7
Newfoundland	–	–	0.1	0.9	5.5	40.8	52.3
PEI	–	47.0	25.4	9.0	13.7	–	5.0
Nova Scotia	–	3.2	19.0	8.2	1.6	0.3	67.8
New Brunswick	–	2.3	16.7	29.5	24.7	0.2	26.7
Quebec	0.1	3.3	4.7	9.5	6.1	0.1	76.3
Ontario	8.9	9.2	12.0	10.9	7.9	4.7	46.4
Manitoba	1.3	19.6	18.9	19.6	17.4	16.2	8.0
Saskatchewan	3.2	18.6	29.9	12.0	27.0	9.3	0.1
Alberta	2.0	10.3	15.6	23.7	28.3	10.0	10.6
BC	0.2	1.3	3.3	7.1	20.5	17.9	49.6
Canada	2.3	8.9	13.8	13.7	18.3	9.9	33.2

		% of Canada's CLI classes by province						
	Total CLI area	1	2	3	4	5	6	7
Newfdland	3.8	–	–	0.0	0.2	1.2	15.7	6.1
PEI	0.3	–	1.6	0.6	0.2	0.2	–	0.1
Nova Scotia	2.8	–	1.0	3.9	1.7	0.2	0.1	5.7
New Brunsk	3.7	–	1.0	4.5	8.0	5.0	0.1	3.0
Quebec	14.7	0.5	5.6	5.0	10.2	4.9	0.1	33.9
Ontario	13.1	51.4	13.6	11.4	10.4	5.7	6.2	18.3
Manitoba	7.0	3.9	15.5	9.6	9.4	6.6	11.4	1.7
Sask'wan	17.1	23.8	35.9	37.0	14.9	25.2	15.9	0.1
Alberta	21.2	18.8	23.5	24.0	36.6	32.8	21.4	6.8
BC	16.2	1.7	2.3	3.9	8.4	18.2	29.2	24.3
	100.0							

Notes

(a) Percentages are rounded to one decimal place
(b) Canada Land Inventory

Source: Canada Land Inventory

flexibility of the agricultural land resource to respond to changing market demands. The distribution of agricultural land by agricultural capability class for Canada's provinces is a striking example of the overall inverse relationship between quality and quantity (Table 2.4). Second, the City's Countryside tends to contain a higher proportion of good quality farmland and a high proportion of the good quality farmland in any national system; Canada again provides a good example of this (Table 2.5).

Table 2.5 The distribution of land by agricultural capability class within an 80 km (50 mile) radius of Canada's Census Metropolitan Areas

Census Metropolitan Areas	CLI classes percentage distribution[a]				
	1–3	4–6	7	Organic	Unclassified
Calgary	49.6	36.8	11.9	0.5	1.4
Chicoutimi-Jonquière	5.0	9.6	84.6	4.3	0.1
Edmonton	64.8	28.8	0.4	5.0	1.0
Halifax	29.0	9.2	60.4	1.4	0.0
Hamilton	77.6	14.2	0.5	2.9	4.9
Kitchener	82.9	10.1	0.6	4.4	1.6
London	86.8	10.2	0.6	1.9	0.5
Montreal	50.0	24.1	19.2	4.1	2.7
Oshawa	63.4	24.1	0.1	8.1	4.2
Ottawa-Hull	33.8	24.9	34.0	6.9	0.4
Quebec	11.2	31.0	51.4	6.0	0.4
Regina	71.4	27.8	0.2	–	0.6
Saint John	10.7	40.7	38.2	0.1	10.0
St Catharines-Niagara	82.7	5.4	0.6	1.0	10.3
St John's	0.2	33.4	49.9	16.8	0.3
Saskatoon	55.1	44.6	0.1	–	0.3
Sudbury	3.2	8.7	86.4	1.7	0.1
Thunder Bay	9.8	27.0	59.7	3.2	0.3
Toronto	71.4	17.2	1.1	5.2	5.2
Vancouver	5.2	12.7	22.0	1.7	58.4
Victoria	9.5	18.3	10.0	0.5	61.62
Windsor	94.3	0.5	3.1	0.6	1.5
Winnipeg	62.4	27.6	1.3	7.9	0.8
Canada	42.5	23.9	25.9	3.6	4.1

Notes
(a) Percentages are rounded to one decimal place
Source: Compiled from Simpson-Lewis (1979)

Given this strong coincidence between cities and land that is well suited for agriculture and the fact that many developers at the edge of cities actually prefer high quality agricultural land because development costs are relatively low (relatively flat and well-drained land), it is not surprising to find high proportions of good quality land being converted to urban uses (Bryant et al., 1982). Between 1966 and 1976, for instance, a period of rapid and unparalleled urban growth in Canada, expansion of the nation's largest cities resulted in the conversion of nearly 150,000 ha of rural land (Table 2.6). Over 60 per cent of this total had a high capability rating for agriculture, and in the case of some cities all of the land converted was well suited for agriculture. From 1966 to 1986, around seventy Canadian cities (populations over 25,000) in this series of studies, 301,000 ha of rural land were converted to urban uses, of which 58 per cent were prime (Classes I to III of the Canada Land Inventory land capability for agriculture classification). Certainly, therefore, in Canada, the conversion of rural land has been disproportionately concentrated in areas with better quality agricultural land (Figure 2.4).

Table 2.6 The urbanisation of rural land in Canada, 1966–76 by agricultural land capability

Land capability	Hectares urbanised	Percentage of total
High (classes 1 to 3) Agricultural Capability	93,000	62.3
Low (classes 4 to 6) Agricultural Capability	32,051	21.5
No (class 7) Agricultural Capability	7,824	5.2
Organic Soils	2,071	1.4
Other Lands	14,450	9.7
Total	149,396	100.0

Source: Abstracted from Warren and Rump (1981, Figure 7)

This pattern is encountered frequently (Best, 1981; Vining et al., 1977). In Scotland, for example, nearly 53 per cent of the 9,998 ha of agricultural land converted to urban uses in the period 1971–79 were prime for agriculture (Department of Agriculture and Fisheries for Scotland, 1982). For the United States, Vining et al. (1977) report that between 1967 and 1975, prime farmland (Soil Capability Classes I and II of the Soil Conservation scheme) was being converted twice as fast as all land and three times as fast as non-prime rural land. Finally, in New Zealand, a country which consistently relies upon agriculture for more than half of its export earnings, it was estimated in 1974 that since European settlement, more than 290,000 ha or about 1 per cent of New Zealand's land area had been urbanised, and 37 per cent of this consisted of 'soils of high value for food production' (Leamy, 1974, 189).

On a general level, therefore, it is true that urban growth in many parts of the world has resulted in the transfer of a substantial amount of good

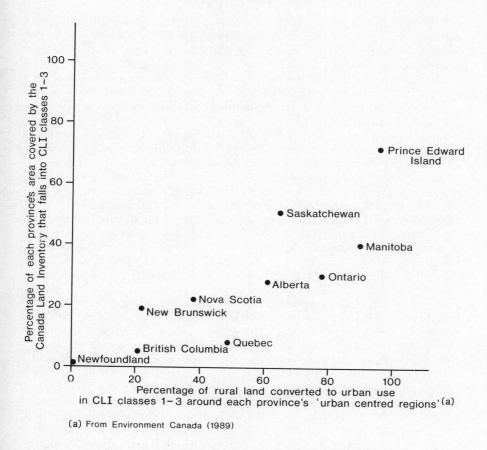

(a) From Environment Canada (1989)

Figure 2.4 Relationship between provincial land capability for agriculture and conversion of prime land, Canda, 1966–86

farmland to urban uses. The term 'prime' land, as used above, is based on soil capability; if climatic considerations are taken into account, they tend to reinforce the conclusions presented so far (Environment Canada, 1989). However, it is not at all clear that in the broader scheme of things, urbanisation of farmland has really led to a major problem yet. Simply determining the absolute amount of land converted, or its quality, tells us very little about the significance of farmland losses. What is needed is information about the consequences of farmland urbanisation *vis-à-vis* the productive capacity of the agricultural system, and for this, it is necessary to investigate further what constitutes a sustainable food production system.

Obviously, one major element of such a system is concerned with the inherent quality as well as the volume of the agricultural resources used in production (Table 2.3). The farmland base is therefore one major ingredient, and changes in the volume of the resource (e.g. through agricultural land conversions) represent one important dimension. It is also important to appreciate changes in the 'quality' of the land resource. This includes improvements in productivity through technological change and, conversely, degradation of the soil resource as a result of erosion and other types of land degradation. These must be taken into account in any long-term evaluation of the sustainability of the whole production system nationally, or – what is even more difficult – internationally.

In addition, it is also important to recognise the role of labour and capital as resources in the maintenance of the sustainability of the system. Notwithstanding increasing concern over the negative impact of farm technology on the long-term productivity of the soil resource for agriculture, it is through changes in the capital inputs in farming, particularly through the use of herbicides, pesticides and fertilisers, that the tremendous increases in agricultural productivity have been achieved during the twentieth century. These increases in productivity have so far more than compensated for any losses in production capacity related to the conversion of agricultural land to urban uses. Ongoing research in the domain of biotechnology applied to agriculture appears to have the potential to extend these productivity advances even further (Lowe et al., 1990).

Another critical dimension of a sustainable food production system is whether it is able to satisfy the needs (WCED, 1987) and, more specifically, the market for food produce. This question of the sufficiency of the food production system is a difficult one (Brklacich, 1989). Are we talking about the ability of one region to meet its own needs, or a country, a group of countries or the world? The difficulty, of course, is that in our increasingly open world economic system, interdependencies are the rule, and it is naïve to assume that any given area should be self-sufficient in food production. Issues relating to the distribution of food as well as production are therefore important (Gittinger et al., 1987). None the less, dealing with the issue of the ability of the production system to meet demands is critical, and various attempts have been made to assess this. It is easier to undertake this for a given country from the perspective of the ability of the system to meet current and future (under various assumptions, of course) demands, and to treat exports as a residual. Various attempts to assess the sufficiency of national agricultural production have been made, varying in the level of sophistication of their methodology.

One assessment of urban encroachment on rural lands in the United States was undertaken by Hart (1976). He concluded that even though 90.9 million acres (36.8 million ha) of land are likely to be urbanised by the year 2000, this amount represents only 4 per cent of the nation's total land area. More importantly for our discussion, he concluded that urbanisation of farmland poses little threat to agricultural production. In the case of urban expansion in England and Wales, Edwards (1969) reached a similar conclusion. In Ontario, Canada, Cocklin (1981), in his study of future urban expansion, concluded that while an estimated 92,000 hectares of Class 1 Equivalent farmland could be converted for urban purposes by the year 2001, 'the intrinsic food producing capacity of the province will be little affected'

(Cocklin, 1981, 104). And in Leamy's study (1974), the high quality land urbanised in New Zealand since European settlement represented only 4 per cent of New Zealand's highly productive soils and did not appear to pose a major threat to the country's productive capacity in farming. None the less, Leamy (1974) suggested that the land involved is not inexhaustible and it is a resource upon which the national economy prospers.

Some of the most comprehensive attempts to assess the significance of future urban expansion for agricultural production have been completed by the Land Evaluation Group, an interdisciplinary team based at the University of Guelph, Ontario. In one study, Smit and Cocklin (1981) estimated the extent of future rural-to-urban land use conversion for each of thirty Ontario counties for the period 1976–2001 for each of four future urban growth scenarios (Table 2.7). The total amount of land estimated for conversion ranged from 68,119 ha under the low population growth scenarios (I and III) to 130,253 ha under the high population growth scenarios (II and IV). The amount of prime agricultural land that would be converted ranged from 31,425 ha under scenario I to 121,515 under scenario IV (Table 2.8). For many observers, the finding that more than 100,000 ha of Canada's best farmland could be converted to urban uses in the last twenty years of the twentieth century represents a serious threat to agricultural production capacity in the province of Ontario. However, even under the 'worst' scenario for those opposed to urban expansion onto good agricultural land, it was estimated that less than 2 per cent of the prime agricultural land base (1976) of Ontario would have been converted over the twenty-five years ending in 2001 (Table 2.9). Thus, on the aggregate level, rural-to-urban land use conversion to the end of the century appears to pose little threat to food production.

It is important, however, to investigate this in relation to food production 'needs' under different scenarios. Therefore, a more analytically elegant

Table 2.7 Future urban growth scenarios for Ontario used by Smit and Cocklin (1981)

Scenario	Scenarios for total provincial population in 2001 (millions)	Location of growth
I	10.0	Growth directed to lower capability soils
II	11.9	Growth directed to lower capability soils
III	10.1	Growth concentrated on higher capability soils
IV	11.9	Growth concentrated on higher capability soils

Source: Compiled from Smit and Cocklin (1981)

Table 2.8 Land requirements from agriculture under different urban growth scenarios, Ontario, 1976–2001

	Scenarios			
	I	II	III	IV
Total land in classes 1–3 (ha)	31,425	80,981	64,386	121,515
Per cent of total	46	62	94	93
Total land in classes 4–6 (ha)	27,137	36,377	2,661	5,145
Per cent of total	40	28	4	4
Total land in class 7 (ha)	9,557	12,895	1,072	3,593
Per cent of total	14	10	2	3
Total land converted (ha)	68,119	130,253	68,119	130,253
Per cent	100	100	100	100

Source: Adapted from Smit and Cocklin (1981).

Table 2.9 Estimated rural-to-urban land conversion (1976–2001) as a percentage of total land in Ontario by agricultural capability

Capability classes	Scenarios			
	I	II	III	IV
1–3	0.5	1.2	1.0	1.9
4–6	0.7	0.9	0.1	0.1
7	0.1	0.2	0.0	0.1

Source: Adapted from Smit and Cocklin (1981).

investigation of the implications of future urban expansion for agricultural production using a mathematical programming model was also undertaken by the same research group (Smit et al., 1983). The researchers assessed the degree to which the achievement of agricultural production targets would be compromised under each of five different scenarios (Table 2.10).

The results show that projected expansions of large cities by themselves present little threat to productive capacity, although under scenario II, it is likely that additional areas would have to be brought into production. These researchers found that the most serious erosion of production capacity would probably come as the result of the expansion of large cities, combined with the growth of smaller places and exurban development. Moreover, the ability of Ontario's agricultural land base to meet future food requirements is more dependent upon land in particular areas being available for production: 'retention for agriculture of lands in the warmer zones and of higher productive capacity, especially for fruits and vegetables, will contribute far more to the ability to meet the specified food needs than the retention of other lands' (Smit et al., 1983, 119).

Table 2.10 Scenarios (Ontario) used by Smit et al. (1983) to assess the implications of future urban expansion on food production

Conditions	Scenarios				
	I	II	III	IV	V
Provincial population in 2001 (millions)	11.0	11.9	←——— 11.0 ———→		
Urban population density in 2001 compared to 1976	no change	20% lower	no change		
Land for expansion of cities with 1976 population > 25,000	83,000	132,000	←——— 83,000 ———→		
Land for expansion of smaller cities and exurban development	none	202,000	←——— none ———→		
Controls on urban expansion	←——— none ———→				high quality farmland protected

Source: Abstracted from Smit at al. (1983, 114)

    This research also suggests that a variety of other options exist for accommodating the land needs from agriculture of non-farm development in the City's Countryside. In addition to the possibility of bringing into cultivation lands in other regions, domestic demands for food could also conceivably be met through reducing exports and/or increasing imports. While this might have negative implications for the balance of trade, it may also strengthen the balance of trade if the uses to which the land is put finally generate, directly or indirectly, more export dollars than the loss in agricultural exports. The types of interrelationships which we have now introduced into the discussion should caution any reader about the dangers of drawing conclusions too rapidly.

    The interdependencies involved in agricultural production for a world market and the complex systems of exchange in which this involves agriculture has indeed been used in the argument for the conservation of good quality farmland (Table 2.3). *One of the lines of argument* deals with the present, and in some cases the future, balance of payments at national level. Some attempts have been made to assess in a fairly direct manner the impact of farmland conversion on export earnings. Stonyer's analysis (1973) for New Zealand followed a series of simple steps:

1. Determine the amount of farmland in each agricultural capability class converted to urban uses (Table 2.11).
2. Express the productivity per unit area of each capability class in either pounds of milk fat (lbMF) or ewe equivalents (EE).

3. Determine the export value of one pound of milk fat and one ewe equivalent.
4. Calculate total lost export earings by summing over the lost earnings over all capability classes.

Table 2.11 Results of Stonyer's technique for expressing the implications of farmland urbanisation in terms of lost export earnings

Agricultural capability class	Acres converted (1973)	Lost production	$ per unit of production	Total ($)
1	3,100	400lb MF	0.58	719,200
2	2,250	300lb MF	0.58	391,500
3	400	2.5 EE	6.8	6,800
4	550	280lb MF	0.58	89,320
5	400	2.3 EE	6.8	6,256
			Total	1,213,076

Source: Adapted from Stonyer (1973)

Using this approach, Stonyer determined that in 1973, New Zealand lost the equivalent of NZ$1.2 million in export earnings as the result of the urbanisation of farmland. Without considering the reliability of the urbanisation estimates (no details are given by the author on the estimating procedures used), two criticisms can be made. First, in terms of the total productive capacity of the agricultural production system, no allowance is made in this type of analysis for any capacity lost through conversion of farmland to urban uses to be compensated elsewhere in the system, e.g. through intensification or through bringing more land into production. Second, in terms of the trade balance, some of the newly urbanised land may be put into uses that either directly or indirectly earn export dollars.

A more comprehensive analytic framework is called for, and it is suggested that scenario-building provides a potentially significant tool. For some countries, like Canada (Lin and Labrosse, 1980), the level of foreign demand is very important for their trade balance. While it is difficult to forecast international demand for a particular country's commodities because of the large number of factors involved, it is possible through scenario analysis to investigate the contribution that the production system can make to international supply once domestic needs have been met. Then, it is possible to examine the conditions under which international demand could absorb the residual amount.

Evaluation of domestic demand is therefore an important component in such an exercise. This involves not only developing population forecasting models with all the assumptions that this entails for such factors as immigration and birth rates, but also making assumptions about consumer tastes. Most studies so far rely implicitly on 'current' consumer preferences. Yet it is known that consumer preferences are not constant. First, they are linked to overall level of socio-economic development (Grigg, 1984). Second, the average food basket even in the developed world has been changing, for example in relation to people's increasing health consciousness and growing concern regarding the residual effects of such things as herbicides, pesticides and growth stimulants in animals (Blunden and Curry, 1988; Lowe et al., 1990; Marsden et al., 1990).

A *second line of concern* deals with the costs involved when interdependencies are really dependencies. 'Imports' of foodstuffs represent one way of compensating for lost 'local' production capacity. However, the argument in relation to agriculture generally and agricultural land in the City's Countryside specifically is that dependencies in terms of imports of foods may become very costly and even catastrophic if the traditional supply lines are broken through some form of disaster or significant change in supply conditions.

Clearly, such concerns were of significance to many of the early proponents of farmland conservation in the United Kingdom, a country which had had to struggle to maintain lifelines for food during World War II. Such anxiety is still expressed as part of the national rationale for protecting farmland in some countries, e.g. in Switzerland with its very small arable land base (Baschung, 1987). More recently, concern over the ability to maintain the food supplies of Paris, with a population in the metropolitan region of roughly 10 million, were reflected in a study of the capacity of the regional agricultural system (the City's Countryside) to respond to the needs of the agglomeration should the traditional supplies (both 'local' and non-local) be reduced as the result of an energy crisis (Philippe and Biancale, 1981). The agricultural sectors in the Paris region which would be most seriously affected by such a change in supply conditions would be fresh fruits and vegetables. For the most part, these are imported from other French regions and from abroad and are therefore very dependent upon transportation linkages, especially since the commercially available storage facilities for fresh produce in the early 1980s was only enough for fifteen days' supply for the agglomeration.

Our discussion of these interdependencies highlights a more general issue. In areas like the Golden Horseshoe of Southern Ontario, the Eastern Seaboard of the United States, or the Greater London Region, there is no question that farmland is being consumed by rapidly advancing urban margins and that farmland in these areas is in very limited supply. However, our assessment of the seriousness of this situation depends on our scale of analysis. Is it important, for example, that the region around Los Angeles meet the food requirements of that city? Or should this be a concern only for the State of California? Given the openness of the system in which agriculture operates, questions such as these underlie the difficulties of developing guidelines for the implementation of sustainable development principles for agricultural land conservation at the local and regional levels. The difficulty suggests we need to devote research to understanding the role that agricultural production in a given region plays in the whole (national, continental) production system.

In addition to the inherent difficulties involved in establishing the significance of conversion of agricultural lands to urban uses, it is important to realise that conversion is not the only process which may be influencing the productive capacity of the agricultural system (Table 2.3). If there are other processes also leading to an erosion of the system's productive capacity, then the issue of the loss of agricultural land to other uses may take on a different perspective.

Two such processes, which have been identified are the result of climatic warming consequent upon the 'greenhouse' effect and the negative impact of modern agricultural practices upon the soil resource for agricultural production (Brklacich et al., 1989). Both are related to the negative effects of human activities upon the environment. The result of climatic warming is difficult to evaluate because of the many unknowns, and even the impact of modern agricultural practices are not easy to pin down because of the variable effects of different biophysical environments for agriculture. None the less, the latter has received increasing attention in several countries (Coote et al., 1981; Dahlberg, 1986, 1988; Munton, 1983; Senate, 1984). The increasing use of pesticides, herbicides and chemical fertilisers in the quest for higher levels of productivity as well as poor manure management systems associated with large-scale animal/poultry enterprises have long been recognised as a major source of pollution in ground-water systems. It is much more recently, however, that concern has become widespread over the effects of poor ploughing practices, monocultural systems involving row crop production, and the use of heavy agricultural machinery on erodibility of the soil, causing compaction of the soil and thus facilitating surface run-off (e.g. Tousaw, 1991). Indeed, the potential relationships between technological change and the evolution of agricultural and rural areas are far-reaching both in the City's Countryside and beyond (for a wide-ranging discussion, see Lowe et al. (1990)).

Where the agricultural system is associated with large open fields such as in the Canadian Prairies and where field consolidation has been extensive such as in parts of the Paris basin following remembrement operations, wind erosion of the soil exacerbates the problem. While it is difficult to put values on the losses involved, a report from the Canadian Senate (Senate, 1984) suggests that the combined impact of these technological and cultural practices could be much more devastating upon the long-term productive potential of the soil resource than the loss of agricultural land to urban and non-farm development. By the same token, the loss of agricultural land to urban use then begins to take on a whole new perspective. To the extent that the current high levels of agricultural productivity have been achieved through intensification, which could be considered fragile in the long term because of the damaging impacts on the soil's productivity, the loss of productive land 'permanently' to urban development is potentially much more significant than previous analyses suggest.

The discussion of the consequences of urban expansion on the sustainability of the food production system has remained inconclusive from a purely quantitative perspective because of (1) the complexities of the processes influencing the productive capacity of the system; (2) our relative lack of knowledge of some of these processes; and (3) the uncertainties inherent in trying to gauge future conditions. This does not mean that inaction is appropriate (see Chapter 5) or that additional research is not warranted. Indeed, partly because of these three factors, one of the more effective ways of addressing the issues is through the development of scenario analysis, in which assumptions about change and future conditions are made explicit (Preston et al., 1987). This approach permits sensitivity analysis to be undertaken and is particularly appropriate for answering 'What if?' questions. Certain types of scenario analysis have already been discussed briefly (e.g. Smit et al., 1983). Another example, which this time dealt with the impact of land degradation (soil erosion, climatic warming) in a part of South-West Ontario, highlighted the importance of considering a variety of changes taken together, rather than simply focusing upon one process, such as land conversion (Brklacich, 1989).

The case of special agricultural lands

Besides the concerns noted earlier about prime agricultural lands generally, there are also special agricultural resource lands which, by definition, are even more limited in supply than prime agricultural lands. Evaluating the impact of urban expansion on these lands is no easier than for prime lands generally, but since they have frequently been singled out for special treatment both in the research literature and from a planning perspective, some brief additional comments are made here.

The case of tender fruit production in Canada offers an excellent example. According to Krueger (1959, 1968, 1977a, 1982), there are really only two areas in Canada where soil and climatic conditions are such that tender fruit (e.g. peaches, plums and cherries) can be commercially grown: the Okanagan Valley in the interior of British Columbia and the Niagara Peninsula in southern Ontario. These areas are limited in their spatial extent and even during periods of peak production are unable to supply the entire Canadian market. Moreover, both areas have long been subject to urban development pressures (Krueger, 1968; Krueger and Maguire, 1984). Given the scarcity of land capable of producing tender fruits in Canada, some would argue that this is sufficient basis alone for restricting the degree to which such land is developed for urban purposes.

The notion of scarcity is developed in the next section. However, the argument outlined earlier regarding the potential replacement of domestic production by imported substitutes in the general case also holds for these 'special' agricultural lands. The key is recognising that if, for whatever reason, these types of lands are valued highly enough to be protected in some way, there are costs involved in this. The ultimate test of any protection is: 1) whether the farmers involved are able to derive a satisfactory standard of living through working the land; and 2) whether society (the municipality, region, province, state, nation) is prepared to shoulder the immediate costs of such a protection and forgo any immediate (short-term) benefits that could be derived from alternative use of the land resource.

Evaluating farmland

Evaluation of farmland is central to much of the debate over the conversion of agricultural lands. Other than referring to some land capability classifications in the discussion above, no systematic consideration has yet been given in this book to the evaluation question. This is the subject of this short section.

It is not surprising that various approaches have been developed to try to evaluate agricultural land on different geographic levels – local, regional and national. A typology of schemes developed by Smit et al. (1987) groups these approaches into four categories: physical quality (categoric), physical quality (parametric), productivity indices and integrated suitability. The progression from the first three to the fourth illustrates well the multidimensional nature of land evaluation for agriculture, and reflects the increasing recognition of the complexity of agricultural land evaluation.

Physical quality schemes (categoric) rate the quality of the land to support agriculture on the basis of the physical properties of the land. Land placed in the highest category possesses no limitations to agricultural production with current technology, while each subsequent category possesses increasing limitations to agriculture (both in number of types of limitations and degree of severity of limitations). Parcels of land assigned to the lowest class in such classification systems are typically incapable of supporting agriculture. Physical quality (categoric) schemes are used in many countries, including Canada, Great Britain and New Zealand, and they have a strong affinity with the Land Capability Classification Scheme developed by the US Department of Agriculture (see Klingebiel and Montgomery, 1961).

Physical quality (parametric) schemes also rate land on the basis of physical attributes, but in these methods parcels of land are rated in terms of how they score on a number of separate physical properties. One of the earliest parametric schemes devised was the Storie Index, developed to rate soils commonly cultivated in California (Storie, 1964). Under this technique, aspects of a parcel's physical attributes including soil profile, surface texture, slope, drainage and acidity are all assessed on a scale of 0 to 100. These separate scores are then multiplied together to yield a single value. The higher this value, the more suitable is the parcel for the particular use for which the method has been calibrated. Other examples of this rating scheme type include the Agroclimatic Resource Index or the ACRI (Williams et al., 1978) and the Climatic Moisture Index (Sly, 1970).

The ACRI has been used in an interesting way to investigate some aspects of how the agricultural production system might 'replace' good quality lands converted for urban development purposes (Environment Canada, 1989). The ACRI takes into account the number of frost-free days, moisture shortages and inadequate summer heat to arrive at an index that ranges from 3.0 for the most suitable lands in south-western Ontario to less than 1.0 in the most northern areas (Williams, 1983). Since there are vast areas of poor agricultural land in Canada, they represent a potential replacement for the capacity lost through land conversion. However, in Environment Canada's study (1989) on land conversion between 1981 and 1986 around Canadian cities, 63 per cent of the roughly 32,000 ha prime land converted was in areas with ACRI values of 2.0 or more. Replacing these lands 'would require more than twice as much land (71,547 ha) of similar soil quality in areas where the ACRI value is 1' (Environment Canada, 1989, 7), not to mention the significantly higher costs involved in such locations.

Productivity indices provide a numerical value based on actual or potential yields. Land uses for which productivity indices are derived are typically narrowly defined. An example of this type of rating scheme is the Canadian Class I Equivalent Productivity Index (EPI) developed by Hoffman (1971). To calculate a Class I EPI for a given land parcel (PI), the percentage of the parcel's total area in each of Canada's Land Inventory Capability Classes (A for Class C) is multiplied by the appropriate yield index (C) and the products summed. The calculation of these measures can be expressed as follows:

$$PI = C \times A$$

Integrated suitability techniques combine information on the physical attributes of land with information on socio-economic characteristics. Schemes in this class take either categoric or parametric forms. An example of this type of rating scheme, developed by Huddlestone and Pease (1979), incorporates four factors: soil suitability, parcel size, surrounding parcel sizes and adjacent land uses. Then, for a given parcel, scores for each criterion are determined, each score is multiplied by an assigned weight and the products are summed.

Despite the popularity of these methods, the notion of identifying land for inclusion in farmland conservation schemes on the basis of suitability alone has been sharply criticised. Several fundamental problems can be identified. In the first place, why should any parcel of land be preserved for a particular use just because it happens to be well suited for that use? Moreover, to what activity should land be devoted if it is found that the parcel is well suited for more than one use?

Some proponents of farmland protection believe that agricultural land needs should be met first, after which all other land use activities can be accommodated. This argument, based on the view that agriculture is the cornerstone of life, neglects the complex reality of land use in the City's Countryside. As we shall see later, land in the City's Countryside serves many different functions. If agriculture is to be given preferred access to this resource in some contexts, then this decision needs to be based on more than some Jeffersonian ideal. It needs to be based on clearly articulated land use goals, along with systematic and rigorous assessment of options and priorities (see Chapter 5). This underlines the need to see agricultural land conversion not only in the context of other processes influencing the capacity of the production system but also in terms of other land uses and functions. For many reasons, the articulation of land use goals and the establishment of land use priorities is a political issue and needs to be dealt with in that arena, on all levels. Information of the relative merits of various land use options is, however, critical to whatever political decision-making exists, and this clearly provides an important role for land use researchers and analysts.

One way that has been suggested for addressing some of the problems associated with nomination of land for inclusion in farmland conservation schemes on the basis of suitability alone is to determine the importance of particular parcels of land for agriculture. This introduces the questions in a manner that highlights some of the significant issues involved in setting goals and priorities. Smit et al. (1987) argue that the importance of land for agriculture can be determined on the basis of three criteria: the suitability of land for agriculture, the uniqueness or scarcity of land for the use in question, and the demand for the products in question. Under such a system, a particular parcel of land is agriculturally important if: (1) it is suitable for agriculture; (2) land with similar attributes is in short supply; and (3) demand exists for the products for which the parcel is well suited.

The concept of resource use flexibility and criticality are closely related to this notion of importance and offer another way of evaluating agricultural land in the City's Countryside (Chapman et al., 1984; Flaherty et al., 1987, 1988) by taking them into consideration explicitly in the context of goals and priorities. Briefly, if a stated land use goal can be easily satisfied, given the resource base of a particular region, then flexibility exists regarding the use of the resource base. If, say, a region's ability to produce a crop far outstrips demand for the crop, then the resource use system would be judged flexible. On the other hand, if the resource base can only just meet the demand for that activity, then the resource use system is considered less flexible and it becomes more critical that land in the region suitable for the activities in question be used for them. To summarise, flexibility provides an indication of the degree to which resource use options exist, whereas criticality is a measure of the degree to which stated goals depend upon particular land types being devoted to particular activities. The relevance of this in the context of 'specialty' agricultural lands is obvious.

In the resource assessment work undertaken by the Land Evaluation Group at the University of Guelph (see for example, Dyer et al., 1982; Smit, 1981; Smit et al., 1984), this was further developed by trying to assess the flexibility and criticality issues using a linear programming model (Chapman et al., 1984). Programming approaches have proved extremely useful in a wide range of location–allocation problems. However, in contrast to the conventional application of programming techniques to resource use problems in which the aim is to identify an optimal solution (see for example, Spaulding and Heady, 1977), the Guelph approach focuses on the set of feasible solutions, or the set of resource use options which satisfy the objective of the model subject to a specified set of conditions. The greater the number of possible solutions, the greater the flexibility of resource use. However, deriving an appropriate measure of criticality which can then be used in resource assessment has proved to be somewhat problematic (Flaherty, Chapman and Smit, 1988).

The introduction of the issues of flexibility and criticality represents a more sophisticated perspective of the evaluation of farmland than the other methods noted above. Even so, this still leaves unanswered the earlier questions of the scale of the region under examination and how to evaluate the appropriateness of different degrees of self-sufficiency in agricultural production. (On what scale can and should the evaluation of agricultural land be undertaken – local, regional, national or continental?) Much more comprehensive procedures are obviously required. One of the most promising approaches, already noted several times earlier in this chapter, is the development of scenario analyses incorporating the effects of various processes that influence the productive capability of the agricultural system (see for example, Brklacich, 1989). An even more comprehensive approach would be to incorporate simultaneously the capabilities of the land resource to support other functions and to respond to goals and priorities for the use of land for these other functions as well. Difficult but significant research questions have to be addressed in scenario building regarding the evaluation of future demands for food, housing, recreation and other activities and the incorporation of technological and environmental changes which have the potential to influence future agricultural production. Despite the difficulties, scenario analysis holds much promise for land use planning in general (cf. Markusse, 1991).

Landscape considerations

Up to now we have focused on the issue of food production and the possible erosion of productive capacity in the City's Countryside resulting from urban expansion as well as from other processes. We have also referred to the capability of the land to support other functions but primarily from an exclusionary perspective, i.e. we have talked of the land being used either for farming, housing, industrial development and the like. However, there are other functions that the agricultural land can perform and support at the same time as it is being farmed. Some of the concern in some countries for the loss of agricultural land to urban and non-farm development has arisen almost as much from losing these other functions as from the loss of the land's productive capacity itself (Table 2.3).

These other functions are related to the various dimensions of the landscape amenity value of farm landscapes, including the role of farming landscapes in providing variation in the landscape composition or the visual and functional composition of natural and human phenomena (Cosgrove, 1984), its open space function in an urban-centred region for both passive and active recreational use by residents, and its role in providing the context in which historic, cultural and natural environment elements are embedded (Bryant, 1986b). Naturally, the relationship is not a simple one. This is illustrated by the increasing conflict between the imperatives of modern agricultural practices and technology, on the one hand, and the difficulties of facilitating multiple use of certain types of agricultural lands for recreation and the conservation of historic, cultural and natural environment elements, on the other.

In some contexts these other values have played a much more important role in the management of agricultural spaces. In the Ile-de-France (Paris) region, for example, Bryant (1986b) points out that the impact of urban growth upon productive agricultural land, although considerable, has never been regarded as a significant issue. Planning efforts have been much more concerned with protecting the integrity of the regional landscape, and the protection of farmland only becomes important because it supports valued open space landscapes.

Similar perspectives are found in other West European countries. In some parts of the Swiss Alps, for instance, agriculture is seen as an integral part of a landscape which is important in terms of recreation and tourism. Many local authorities have established programmes to support mountain agriculture, offsetting the high costs of production with subsidies on grain as well as income supplements (Price, 1981). As another example, the value of particular forms of farming in maintaining certain types of countryside has long been recognised in the United Kingdom, even to the extent of farmers receiving payment to continue with certain farming practices (Blunden and Curry, 1988). Even in North America, examples of this concern can be found in the area near Philadelphia, Pennsylvania. For instance, attempts have been made by the French and Pickering Creek Trust to preserve historical sites and open space in the face of urbanisation (Morris, 1981).

The multiplicity of resource demands in the City's Countryside

It is because of the multiplicity of demands that are made on land in the City's Countryside, either for exclusive uses or for multiple uses, that effective planning and management within the City's Countryside are so important. We have probed some of the underlying questions regarding the evaluation of farmland in the City's Countryside in this chapter. There appear to be no simple answers because

1. several processes operate simultaneously to influence the capacity of the agricultural food production system;
2. the question of the capacity of the system depends as well on the labour and capital inputs in farming, how they are organised together with the land into different farming systems and how the production is distributed to the market and the consumer; and
3. the sufficiency of the production system also depends upon the shape of future demands to be placed on the system, influenced by the changing composition of the 'food basket' and demographic structures.

We are not able to tackle all of these dimensions in this book, but before we deal more extensively with the issues of public intervention in the planning and management of agricultural land and resources in the City's Countryside (Chapter 5), we shall turn our attention to the systems of exchange, especially the different market structures, which animate farming in the City's Countryside (Chapter 3) and the role of the farmer as decision-maker (Chapter 4).

3
The market: systems and structures of exchange

Introduction

In Chapter 1, agriculture and the individual farm units were seen as being tied into the broader environment and society through various systems of exchange. These systems of exchange link decision-making units both within a given scale of analysis as well as between different scales. Systems of exchange involve two major types, i.e. market-regulated exchanges (e.g. for farm inputs or supplies, capital, agricultural commodities, hired agricultural labour) and non-market exchanges (e.g. for information, and social interaction between farmers or between farmers and non-farmers).

All systems of exchange comprise two fundamental components. First, there are the transactions themselves, involving the objects (goods, supplies, services) and ideas moving within the particular system of exchange; when part of the market system of exchange, these have a monetary value. Second, the exchanges can take place in a network; depending upon the particular system of exchange, this may include the physical infrastructure of communications and social networks as well as the points of contact and exchange (nodes). The nodes, again depending upon the specific system, include the individual farm units, farm input suppliers and food processors, produce collection points, markets, sources of information, and so on.

Involvement of farmers and their families in these systems of exchange is influenced by individual and family factors. These factors as well as the socio-economic organisation of the farm unit also act as a filter to the stimuli and pressures experienced by the individual farm unit and family.

Both types of systems of exchange, market and non-market, can be influenced by public sector or government intervention. We shall comment upon some of these interventions here, but shall leave consideration of the role and nature of public intervention in the agricultural land conservation domain until Chapter 5. In this chapter, our focus is on those systems of exchange which are fundamental to the agricultural production component, although we shall make some limited comments about social interaction.

Markets for agricultural produce

Once agricultural production moves beyond the subsistence mode of organisation, it is the marketing of agricultural produce which provides the *raison d'être* of the farm as a production unit and which provides one of the most important ties between the farm, and agriculture generally, and the broader socio-economic system.

Agricultural production in the City's Countryside and urban markets

The location of agricultural production in relation to urban markets is one of the longest-standing relationships examined by agricultural geographers. It immediately conjures up images of Thunian concentric agricultural land use rings and Thunian analysis (Hall, 1966). The market, represented by the urban centre as the market or point of exchange of farm produce, determines price of produce through the interaction of supply and demand, and this in turn is translated into price signals which are acted upon by the individual farmers in making their enterprise allocation decisions. The costs of transporting farm produce are singled out for special attention in the initial formulation of the Thunian framework. It is the interplay of transport costs, production costs and sales revenue which determines the competitive edge of different enterprises and farming systems, and which leads to the familiar concentric ring formulation of agricultural land use as well as the declining pattern of intensity of production for a given enterprise as distance away from the market increases (Bryant, 1974; Hall, 1966).

Naturally, once the assumptions regarding homogeneity of the land and the closed regional system of exchange in agricultural produce are relaxed, the concentric ring structure breaks down. This does not mean that Thunian principles of competition for land between different uses and farming systems do not still apply (McDonald, 1974). Rather, other factors are recognised as influencing the comparative advantages of different regions and subregions with their different physical and economic productivity relationships. Access to major urban markets, therefore, cannot be seen as simply benefiting only those areas adjacent to the urban centres. As transportation systems have improved (see below), the geographic reach of major markets has increased for a greater variety of farm produce. Despite such technological advances, however, there are still advantages that are enjoyed by some enterprises close to a major urban centre. Many horticultural crops, for example, which are marketed fresh need to reach consumers within days, and sometimes within hours, because of their high degree of perishability (Jumper, 1974). Furthermore, even though tremendous advances have been made in transportation technology, e.g. controlled atmosphere trucks, the cost of transporting many horticultural products remains a major determinant of business profitability (Janick, 1979).

Major urban markets still represent a significant concentration of demand, both in terms of the number of consumers and their generally higher incomes. It is in the expansion of the regional city within the City's Countryside that we expect to find the greatest development of some of the 'new needs' of post-industrial society which have a direct effect upon agricultural production, e.g. demands for fresh, high quality produce and for horticultural products. Furthermore, the key distribution facilities and networks are often anchored in the main centres, as Moran (1987) points out for New Zealand's fresh vegetable production. Sometimes, the control exerted by wholesalers is such as to make it difficult for producers to enter production near smaller market centres.

Both past and present advantages of the near-urban location are reflected in some of the familiar concentrations of certain types of farm enterprises in the City's Countryside (Rutherford, 1966). There are many specialised and intensive agricultural areas that developed to serve the adjacent urban market. Classic examples include the orchard fruit areas north and west of Paris, France, which have their origins in serving the luxury fruit demands of the urban bourgeoisie and aristocracy (AREEAR, 1976; Préfecture de la Région d'Ile-de-France, 1988). These areas developed under technological and market conditions that conferred upon them a relative geographic monopoly. In the United States, Lawrence (1988) shows how the sales value of horticultural specialty crops and vegetables increased significantly in his study of 15 US metropolitan counties from 1949 to 1982 (Table 3.1). Other examples include the varied intensive agricultural production near Auckland, New Zealand (Moran, 1979), the classic cases of dairying near Los Angeles (Gregor, 1963, 1982), peach and other fruit production in the Niagara Fruit Belt of the Niagara Peninsula in south-western Ontario (e.g. Krueger, 1968, 1978), and the varied market-oriented agricultural structure of the Sydney region, Australia (Crabb, 1984; Rutherford, 1966). Other factors such as excellent quality land and climatic conditions have contributed to some of these specialised concentrations, and sometimes the benefits of proximity have been reinforced, so to speak, through public intervention with such approaches as the milk shed (e.g. Codrington, 1979). Proximity to market has been an important factor, however, in all of these examples.

For some farm products, proximity is reinforced because of the high transport costs that are associated with a bulky, low value and fairly perishable product. In North America, a very good example is that of turf production. In those regions where turf production has developed, it exists largely to satisfy an urban demand for an 'instant environment' in the growing suburban areas (Weed, Trees and Turf, 1971). Turf is relatively bulky and low in value and cannot be stored long without deterioration,

Table 3.1 Agricultural production changes in selected US metropolitan counties,[a] 1949–82

	1949–82 % change in		
	Area of crops or number of animals on census farms	Sales values	Intensity of production[b]
Horticultural specialties	4	138	196
Field crops	-38	58	N/A
Other crops	N/A	N/A	234
Vegetables	-59	22	233
Fruits and nuts	-67	-42	152
Poultry	-73	-66	153
Dairy	-86	-73	176

Notes

(a) The counties and their respective metropolitan centres are: Queens, Nassau, Suffolk (New York), Los Angeles (Los Angeles), Cook, DuPage, Kane, Lake (Chicago), Dallas, Tarrant (Dallas-Ft. Worth), Baltimore City and County (Baltimore), Multnomak, Clackamas, Washington (Portland), Polk (Des Moines).
(b) Based on the 1967 dollar sales per unit area harvested or per producing animal, and based on the ratio between intensity figures for the 1978–82 period and the 1949–54 period.

Source: Compiled and calculated from Lawrence (1988, Tables 1-4)

especially in hot, dry weather, so that it is rare for it to be transported more than eighty to one hundred kilometres (fifty to sixty miles).

In addition, the size of the urban market in some regions has encouraged very large-scale production systems, e.g. dairying, beef production, large-scale nurseries and horticultural centres, and intensive poultry and pig farms. It is interesting in this respect to note Wong's analysis of agriculture around Canada's metropolitan centres (Wong, 1983). Although participation in such enterprises as poultry and pig production was widespread throughout both metropolitan and non-metropolitan regions alike, the larger-scale operations were heavily concentrated in the major metropolitan regions. The market effect of a major urban agglomeration is compounded by other factors in these regions such as land and labour costs, but the market effect is important for such operations because it permits producers to capitalise on the economies of large-scale production in some enterprises. It is not surprising then that in some studies of intense urban fringe locations, such as Gregor's (1991) analysis of Orange County, California, substantial increases continue to be registered in the value of agricultural production despite significant absolute declines in the land base and in the areas of land devoted to different enterprises and the numbers of livestock. This is supported by several of Lawrence's (1988) observations, by Gregor's (1988) general analysis of southern California (Table 3.2) and by Crabb's (1984) comments on the Sydney metropolitan region (Table 3.3). As in many other metropolitan areas, these data for the Sydney area also underline the diversity of the farm production systems in the City's Countryside – and in this area, despite what

Table 3.2 Changing patterns of agricultural intensity, Southern California, 1950–80

Counties	1980 % urban	% change 1950–80		
		population	value of crops sold/unit area harvested	value produced per farm[a]
Los Angeles	98.9	80.1	469.8	363.7
Orange	99.7	793.8	1039.3	1373.2
San Diego	93.2	234.4	699.1	468.5
San Bernadino	90.1	217.8	72.1	1133.5
Riverside	82.4	290.0	237.9	569.7
Ventura	94.6	361.6	341.8	256.5
Santa Barbara	90.8	204.1	308.1	153.8
California	91.3	123.6	204.8	449.5

Notes

(a) Value of products produced on farm minus value of products bought in but produced on other farms.

Source: Compiled from various tables in Gregor (1988)

Table 3.3 Farms reporting selected agricultural produce, Sydney Statistical Division, 1981–2

Total no. of agricultural holdings 3,145
% of farms reporting:

Nurseries/flowers	11.7	Mushrooms	1.5
Cultivated turf	1.4	Sown pasture & grasses	14.2
Citrus	7.4	Total hay crops	2.1
Pome fruits	2.1	All crops (excl. pasture & grasses)	61.9
Stone & other orchard fruit	11.8	Sheep & lambs	1.5
Strawberries	1.2	Dairy cattle	15.8
All fruit (incl. grapes)	25.5	Beef cattle	26.4
Potatoes	2.4	Pigs	4.9
Vegetables (human consumption)	24.9	Goats	2.8
		Horses	16.0

Source: Compiled and calculated from Crabb (1984, 271-1)

has been characterised as a region with a short supply of fertile agricultural land (Crabb, 1984; Department of Environment and Planning (Sydney), 1984; Rutherford, 1966).

Transportation of farm produce and agriculture in the City's Countryside

Transportation conditions have much to do with the relative spatial monopoly once enjoyed by the intensive and specialised agricultural areas located near many of the larger European centres. In the days of the horse and cart at the end of the nineteenth century, it was not uncommon for farmers and their families in the vegetable-growing areas in the Seine valley to the west of Paris (e.g. St Germaine-en-Laye) to have to spend a whole day taking their produce to Les Halles in the centre of Paris (Pédelaborde, 1961). Subsequent changes in transportation technology have dramatically altered the competitive and market environment for these areas. This is also true, however, for some of the intensive agricultural areas that developed more recently, e.g. the Niagara Fruit Belt in southern Ontario. Here, one of the problems facing fruit production for a long time has been the competition from regions such as California and Mexico where production costs are considerably lower (Reeds, 1969). This competition has been made possible by developments in refrigerated transportation as well as storage technology.

In Western Europe, improvements in transportation technology from the end of the nineteenth century (rail transportation, and subsequently truck transportation) led to severe competition for some agricultural enterprises and zones near the larger conurbations. Vegetable production in the Paris region suffered from competition from the Loire valley as a result of truck transportation, facilitated by the rapidly evolving system of high-speed motorways. On the other hand, Paris region producers who wished to serve the same wholesale market have had to contend with traffic congestion on the local and regional roads. Fruit production in the Paris region has also suffered from competition from the Mediterranean areas. The magnitude of the impact of this competition, and, in a sense, of the forgone opportunities for local producers in view of the tremendous expansion of the market in the Paris urban area, is shown by the fact that in 1895 the Paris region producers supplied about 81 per cent of the region's needs in fresh fruit and vegetables by weight, compared with only 46 per cent in 1950 (Bryant, 1984a). Increasing competition has also become a fact of life with the lowering of trade restrictions within the European Community. Technology and institutional change have clearly altered these market systems of exchange in major ways.

In view of this, one might question why some of these intensive agricultural areas manage to remain in production at all. We discuss below the marketing advantages of a near-urban location. In addition to these factors, there is also a seasonality factor, which has meant that because of differences between production seasons of different producing regions, it may still be possible for local producers to enjoy some spatial monopolistic advantages at some times of the year. Given price variations over the year, developments in storage technology especially for fresh fruit since the 1960s has allowed some local producers to take advantage of higher price periods by spreading out their harvest. However, this same technology can also be used by producers in other regions to compete more effectively with local production.

The relative decline in transport costs has therefore also contributed to the changing geography of agricultural production on a broad scale. Logically, this should mean that the location of agricultural production should reflect more and more the impact of other factors. In fact, this has been suggested both in some of the agricultural geography texts (e.g. Grigg, 1984; Morgan and Munton, 1971) and in some macro-level research. In Canada, Wong (1983) traces the changing pattern of regional specialisation from 1941 to 1976 in both the metropolitan and adjacent non-metropolitan or broad hinterland regions. Although there were important differences between the agriculture in these two types of region, there were also striking similarities in aggregate agricultural structure and change (Bryant and Wong, 1986; Wong, 1983).

The implications of this are twofold. First, it means that some of the changes in agriculture in the City's Countryside reflect these broad changes in technology and specialisation. Second, it implies that some of the problems faced by agriculture in the City's Countryside may arise from the macro-scale processes affecting interregional and international competition. Not all of the agriculture in the City's Countryside is, of course, influenced to the same degree by such changes, because some farm sectors and farms participate in very different systems of exchange from the marketing perspective. In some of these, the role of transportation and accessibility continues to play a very important role.

Systems of exchange and agricultural marketing in the City's Countryside

The marketing of agricultural produce is complex because there are many different sequences that farm produce can take from the farm production system to the final consumer (Figure 3.1) (Moran, 1987). The complexity arises in the first place because of the organisation of farm production and the nature of consumption of farm produce. Notwithstanding the existence of significant economies of scale in production in some agricultural sectors and the development or infiltration of various corporate and finance capitals in agriculture (Munton, 1985), farm production continues to be dominated by a large number of individual producers.

Similarly, there are an even larger number of individual consumers of farm produce. It is therefore not surprising that various structures and arrangements have been developed

1. to give the farm producer more bargaining power and weight in the sale of farm produce (e.g. produce marketing boards, co-operative selling groups);
2. to facilitate the assembly, grading, preparation and packaging of farm produce (e.g. co-operative structures, wholesalers, the food preparation industry); and
3. to facilitate the distribution of farm produce (e.g. wholesalers, retail operations).

Another development, of course, that some farmers have used to circumvent intermediaries in the marketing process is to become involved in direct marketing.

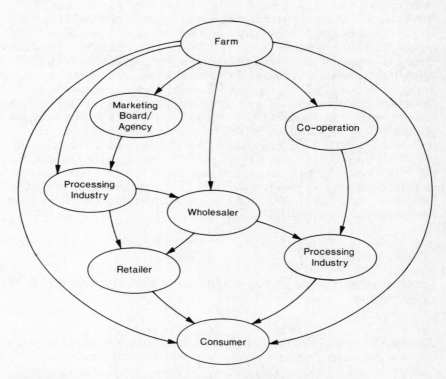

Figure 3.1 Major paths in the marketing of farm produce

Generally speaking, when we think of the marketing of farm produce, we tend not to think of farmers being directly involved in either the processing and packaging of the farm produce. The farmer is seen as relatively distant from the consumer. The farmer is seen as having little control over price, even where there is a marketing board. The focus of attention is on the relatively concentrated or oligopolistic structures associated with the food processing industry and the retail system.

However, in the City's Countryside, even though there is always a significant proportion of the agriculture caught up in systems of exchange which involve the trading of agricultural produce interregionally and even internationally, an environment exists in which part of the farm production sector has been able to develop a much closer link with the sale, and even processing in some instances, of farm produce. This is partly because of the concentration of certain types of large-scale consumers of farm produce in metropolitan regions; partly because of the concentration of some of the 'new needs' associated with post-industrial society there and partly because of the opportunities for 'short-circuiting' the complex links in the marketing chain between farm and consumer which we commonly associate with agricultural marketing (Johnson et al., 1987; Laureau, 1983; Rickard, 1991a; Smith, 1987; Thomson, 1981).

Hence, there are opportunities in the City's Countryside for developing direct farmer to consumer marketing channels (Moran, 1979), including large volume sales direct to supermarkets, factories, schools and hospitals (where there are facilities for preparing meals for workers, students and patients), as well as direct sales to the individual consumer (Laureau, 1983, 1984; Rickard, 1991a). For some farmers, selling direct to the individual consumer has long been a part of the weekly round of the farm family. In Western Europe, markets exist in many villages, towns, suburbs and cities, where farmers and their families have long stood side by side with retailers in selling their produce. Such traditions go back centuries in Western Europe, and despite the competition from supermarket chains, many continue to thrive because of many consumers' desires for perceived freshness and quality of produce. In addition, many of the 'farm' markets that have developed in suburban, or rural and small town locations near urban areas have capitalised upon the urbanites' desires to return to a more 'rural' setting, where they can rub shoulders with 'real' people and enjoy the hustle and bustle of a 'rural' market. In many areas, they also become part of the tourist attractions and are often marketed in those terms. Of course, this phenomenon is not unique to Western Europe, and farmers' markets can be found throughout North America responding to the same sets of 'needs'.

More interesting in the context of the City's Countryside are the opportunities for direct sales to consumers on the farm itself. Specialised horticultural and nursery operations often have a direct sale outlet or garden centre attached to them. Accessibility considerations are important here, although they are expressed in terms of accessibility and time considerations for the customers more than in terms of perishability constraints of the produce. These intensive horticultural activities are a 'natural' for the City's Countryside, and are often found in high traffic zones and not infrequently in clusters (Laureau, 1983).

Other forms of direct sale on the farm itself are seen in the farm shop, a fairly common feature in the United Kingdom, and more generally in the proliferation of various types of 'pick-your-own' ventures, especially concentrated into the most urbanised regions such as the South-East and West Midlands areas of the United Kingdom (Bowler, 1981a,b). Rickard (1991a) also notes a concentration of direct marketers in Connecticut in the urban fringes of Hartford and New Britain, although they were fairly spread out too. What is interesting about the 'pick-your-own' operations is that they can be adapted to a wide variety of farm enterprises and integrated with a wide range of types of socio-economic organisation of production, e.g. the small-scale family and 'artisanal' type of operation and the large scale, capitalistic operation. Obviously, considerations of accessibility and time constraints for the customer are critical, especially in terms of accessibility to a large market for the larger 'pick-your-own'. They are developed not only in major metropolitan regions, but can be found wherever the level of demand is large enough, and this includes some of the holiday resorts e.g. Prince Edward Island in Canada, and Marlborough Sounds in New Zealand's South Island.

'Pick-your-own' farming has developed extensively in North America and Western Europe (Laureau, 1983, 1984). In major metropolitan regions, they can be seen as a response to three sets of factors:

1. the high cost and scarcity of agricultural labour, even casual agricultural labour, near major metropolitan centres;
2. the farmer's perception that a larger portion of the consumer dollar spent on food ends up in the farmer's pocket, because the intermediaries in the selling chain have been eliminated; and
3. the urban dwellers' desires for fresh, high quality produce at a reasonable (perceived, if not always real!) price, combined for some people with a desire for a 'rural', open air experience.

Laureau (1983) reports on an analysis of 250 questionnaires administered at three different times in the week (mid-week, Friday/Saturday and Sunday) to a sample of customers of a large pick-your-own operation north-west of Paris. He found that mid-week customers were primarily attracted to the pick-your-own by price and secondly by produce quality, the Friday/Saturday customers by price and quality simultaneously, and the Sunday customers by a combination of the 'outing' itself and the price of produce. One of the factors underlying the quest for quality produce is for fresh produce that the consumer can process and/or freeze to conserve for later consumption. This is again partly a reflection of freezer technology and partly of a changing life-style.

'Pick-your-own' operations have been developed in a wide variety of enterprises, especially intensive fruit and vegetable production (see, e.g., Table 3.4) (Bowler, 1981a; Crabb, 1984; Rickard, 1991a) Frequently encountered products are raspberries and strawberries in terms of fruit, and beans in terms of vegetables. However, the variety is enormous, and in some areas there is a sufficiently large number of farms and different product types involved that pick-your-own directories and calendars are produced (e.g. FSAP, 1982; Ontario Ministry of Agriculture and Food, 1984,) and some of the larger individual operations with a variety of produce also produce calendars for the customer (see Figure 3.2).

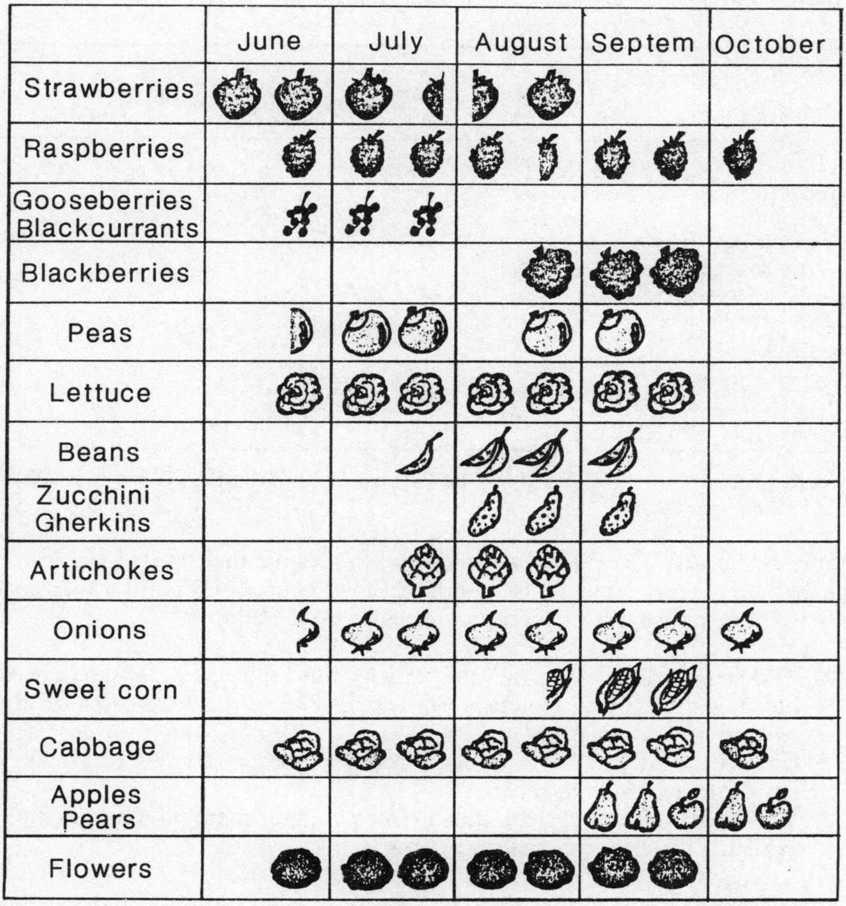

THE MARKET 65

	June	July	August	Septem	October
Strawberries	🍓🍓	🍓🍓	🍓		
Raspberries		🍇🍇🍇	🍇🍇	🍇🍇🍇	🍇
Gooseberries Blackcurrants	🍒	🍒🍒			
Blackberries			🫐	🫐🫐	
Peas	🫛	🫛🫛	🫛	🫛	
Lettuce	🥬🥬	🥬🥬🥬	🥬🥬	🥬🥬	
Beans		🫘	🫘🫘🫘	🫘	
Zucchini Gherkins			🥒🥒	🥒	
Artichokes			🌿🌿🌿		
Onions	🧅	🧅🧅🧅	🧅🧅🧅	🧅🧅🧅	🧅
Sweet corn			🌽	🌽🌽	
Cabbage	🥬🥬🥬	🥬🥬🥬	🥬🥬	🥬🥬	🥬
Apples Pears				🍐🍎🍐🍎	
Flowers	🌺	🌺🌺🌺	🌺🌺🌺	🌺🌺	

Variations in the dates can arise because of changes in weather patterns

Figure 3.2 Pick-your-own calendar: an example from a large PYO northwest of Paris

Pick-your-own is not confined either to crops that customers pick off the plants or pull from the ground themselves. This can have disadvantages, especially when the crop matures over a period of time, because of wastage on the part of customers who may be unaware of how to select the product and because some plants are simply more fragile than others. Furthermore,

Table 3.4 Produce types in direct selling in Connecticut, 1989

Produce	% of Connecticut direct marketers with:			Total no.
	PYO	Stands	PYO & stands	
Berries	49	2	1	36
Tree fruit	13	8	6	24
Vegetables	21	46	8	75
Berries & tree fruit	7	3	7	16
Berries & vegetables	9	4	26	37
Tree fruit & vegetables	–	12	11	25
Berries/tree fruit/ vegetables	1	25	41	71
	100	100	100	
	(68)	(116)	(100)	

Source: Calculated from Rickard (1991a, 84), based upon the 1989 *Connecticut Direct Marketing Directory.*

some of the customers may be turned away – despite their quest for a 'rural' or 'back-to-nature' experience – because of concern about 'getting their feet dirty'. In response to this, various strategies can be adopted:

1. The farmer picks the crop, e.g. carrots and leeks, on a daily basis and bunches them up to be left in piles at the end of the crop rows for the customer to select from.
2. Grassed paths can be developed between the crop rows to facilitate customer access to the crop.
3. The farmer picks the crop, bunches or packages them, and makes them available for customer selection on a stand.

In the last case, we are only a short way from the farm shop, when the farmer may even begin to buy in produce from other farmers or wholesalers to augment the variety of the produce for sale. The pick-your-own philosophy can also be extended to 'dig-your-own' (e.g. shrubs) and 'chop-your-own' (e.g. Christmas trees) (Rickard, 1991a).

Finally, farmers have not been the only ones to capitalise on evolving consumer preferences for fresh and wholesome foods demanded by a clientele that appears to be more and more concerned about health. Supermarkets in the 1980s have begun to cash in on this through the sale of 'environmentally friendly' products: this is being reflected in the sale of organically produced fresh farm produce in supermarkets (COG, 1989).

The pick-your-own phenomenon is therefore closely related to the new needs that are being expressed more and more as our society has become increasingly urbanised. These direct sales outlets provide access to freshness and 'genuine' farm produce, but they also provide, for some of the clientele at least, a recreational experience. Some farmers have capitalised on this by developing picnic facilities as an adjunct to their sales outlet. What better way to spend an afternoon with the family than to be in the countryside with a picnic, part of which you have picked yourself? And even better if there is cream for sale to go with the strawberries!

Most of the products that we have mentioned so far in relation to direct sale have been products that characteristically are not regulated in any substantial way. We shall return to the implications of this in terms of entrepreneurial activity in farming in the City's Countryside in Chapter 4.

Some farmers may become involved in a minimal amount of processing of their produce before selling direct to the public, a stage that 'adds value' to farm production at the farm level. Examples of this have been encountered in Bryant's fieldwork in the Paris region, e.g. processing of milk into butter and yoghurt, and in Bryant and Johnston's fieldwork around Toronto, Southern Ontario, e.g. using a small portion of the fruit production of a farm in making baked goods. It is even more common for farmers to undertake some packaging with a minimal amount of 'processing' of the produce, e.g. grading, washing and bagging potatoes, carrots and apples.

Of course, there are other opportunities for some farms in the City's Countryside which are linked to the recreational demands that emanate from the urban complex (Ilbery, 1988a, b). Some farmers perform non-traditional agricultural services, e.g. riding stables and livery stables for city-dwellers' horses, although other people besides farmers move into this type of operation. For instance, in the Chevreuse Valley south of Paris, most of the riding and livery stables are owned and managed by people who have never been full-time farmers (Benmoussa et al., 1984). Yet they utilise pasture land and, in some cases, may even grow some of their own feed. Clearly, 'horsiculture' becomes part of farming in the City's Countryside. It is simply part of a very different system of exchange and involves different types of socio-economic organisation of production, but it is a characteristic of farming areas in the City's Countryside which has been reported in several countries, including France (Benmoussa et al., 1984), the United Kingdom (Thomson, 1981) and the United States (Irving, 1966).

In Wong's (1983) analysis of changing agricultural structures and patterns around Canada's metropolitan centres, 1941 to 1976, the proportion of census farms reporting horses declined up to the early 1960s, and then began to increase substantially again. This is in contrast to the proportion of census farms reporting other livestock such as beef cattle, dairy cows, pigs, and hens and chickens which all experienced a continued decline. Punter (1976) commented on the emergence of 'horsiculture' in King Township, north of Toronto, confirming this pattern.

Finally, some farmers perform non-agricultural services, the demand for some of which can be expected to be larger and more intense in the City's Countryside. This includes renting out obsolete barn space for storage of recreational vehicles (e.g. boats, dormobiles, infrequently used custom-built cars, and in parts of North America, snowmobiles), and even the preparation of allotment gardens for rental by nearby urban residents (Bryant, 1984a), a form of diversification that Gilg (1990) has referred to as 'structural diversification'. An extension of the link between pick-your-own and the 'Sunday afternoon outing' is the occasional development of a 'complex', usually of different entrepreneurs - both farm and non-farm - involving such activities as PYO, antiques, garden centres and the like in a cluster (e.g. Pinault, 1983).

Markets for produce and markets for agricultural inputs

Agricultural prices are not only a function of the supply of and demand for agricultural produce, but are also influenced by public intervention in the market-place, often in the form of price supports or subsidies to farm producers. Changes in prices reflect changes in supply conditions (e.g. shortfalls in production, changing volume of competitors' imports), changes in consumer demand (e.g. changes in preferences, growth in local market) and changes in government policy (e.g. regarding level of price support). They provide signals to farmers to review their production systems, specifically in terms of their enterprise structure.

The extent to which farmers respond to product price changes depends partly upon the flexibility of the farm operation (i.e. How specialised are the resources on the farm in terms of enterprise mix? How does the level of indebtedness influence choice? (Johnston, 1990)), the flexibility of the farmers (i.e. How willing are they to contemplate change?) and the socio-economic mode of organisation of the farm (e.g. Does the farm operate on a purely capitalistic basis, does it function as a combined production-living unit or family farm, or is it cushioned from some of the vagaries of the market-place because it is run by a hobby farmer?). In relation to the socio-economic organisation of the farm, on the one hand the farmer's response is partly conditioned by the relationship between the farm business and the farm household. If there are significant sources of non-farm income, either from employment or from investment, this may introduce greater stability into the production system. On the other hand, a farm that is run on more capitalistic lines may be more likely to respond to price changes in order to maximise income from farming, whether or not other income sources exist in the farm household.

In market-oriented farm production, changes in product prices will have an impact sooner or later on the farmer's decisions regarding enterprise mix, and therefore resource allocation and the farm's input structure. This introduces an important concept that we need to draw upon for the following discussion on farm input markets, viz. the demand for agricultural inputs (land, labour, capital) as a derived demand, and specifically as a function of the levels of return that can be expected from the sale of the products from utilising those inputs. It is therefore useful to talk of the 'use' value of a given agricultural input, e.g. agricultural land, which is the value that can be ascribed to an input given its productivity. We can calculate this for agricultural land, for instance, by capitalising the net returns for the land under a given system of agricultural production. This value must be distinguished from the actual market value of the input in question. Market value reflects not only the 'use' value of that input in a given use but also the range of possible uses or functions to which the input can be put (Found, 1971).

Finally, how farm inputs are 'valued' by farmers under different socio-economic modes of organisation can be important in influencing decisions to introduce changes on the farm. If certain inputs which are owned by the farmer and/or farm family in the family farm mode of operation are not valued at their market value, there is a tendency for those inputs to be utilised less efficiently (economically speaking) than if a 'true' imputed economic value were assigned to them or if they involved a real cash cost defined by the market (Malassis, 1958; Special Committee on Farm Income, 1969). The latter would be the case for a capitalistic farm which rented its land base and hired its labour force. The importance of this is seen in the increased involvement of farm family labour – the commodification of family labour – in the broader labour market through opportunities for off-farm employment. Where this phenomenon exists, either through actual or potential off-farm employment of family members, especially farm women (Moran et al., 1989), the result leads to a re-evaluation of the value of family labour with all that that entails for reallocation of resource use on the farm.

Markets for farm inputs

Farms are also tied into systems of exchange involving agricultural inputs: farm labour, capital (fixed and operating) and land. Characteristically, the result of the operation of the various systems of exchange leads to a monetary value being placed upon a particular input. However, the monetary or market value may reflect alternative uses or functions for the input. The value of agricultural land and labour may reflect their alternative use values in non-agricultural uses. It may also reflect non-economic or non-production related functions of the input such as the value of agricultural land to a hobby farmer, who may be as interested in farming for the pleasure to be derived from it as in the actual production value.

The various systems of exchange or markets in agricultural inputs interact simply because the same physical input may have several different uses or functions. The interpenetration of the different systems of exchange is therefore affected by information transfer from one system of exchange to another, e.g. information on prices of products, input prices, and alternative opportunities for inputs. In the City's Countryside, the various systems of exchange involving potential and actual agricultural inputs are likely to be more complex in their interrelationships because of the greater range of alternative opportunities and uses, and because of the greater availability of and accessibility to information. It is not surprising that agriculture near the urban-industrial complex has often been seen as dynamic because of the greater density of stimuli and greater speed with which ideas and values diffuse there (e.g. Pautard, 1965). Similarly, we can expect the forms of socio-economic organisation of production to be more varied there, because of the greater likelihood of hybrid forms developing as certain elements, e.g. values and ideas about what constitutes the 'good life', diffuse across different socio-economic structures of production.

Labour

GENERAL CONSIDERATIONS. The human input into agricultural production is the most difficult to conceptualise because of the variety of human resources (e.g. farm family versus hired labour, labour as opposed to management input), the different modes of socio-economic organisation of production (e.g. full-time capitalistic units, full-time 'family' operations, and various types of part-time and hobby farms) and their associated values, and the variety of systems of exchange in which human resources are involved. These span both economic production-oriented systems of exchange and non-economic social and cultural systems of exchange.

Farm production, as noted in Chapter 1, is pursued for several reasons, the relative importance of which varies from one socio-economic mode of production to another. We can identify three broad reasons for the purposes of our discussion:

1. production to provide food directly for the farm family;
2. production to provide a surplus over and above family needs for exchange in the market-place which then permits the purchase of other goods and services needed by the family; and
3. involvement in farm production because of the non-economic values associated with farming: these include the value of living and working together as a family unit as well as the pleasure gained from working in harmony with the land.

Closely associated with the last point is a strong desire to pass the family farm on to the next generation.

As the dominant socio-economic modes of production have changed over time, from subsistence agriculture through to industrialised, capitalistic agriculture, so have the dominant sets of values associated with involvement in farming. With the development of the industrial sector, and industrialisation and urbanisation generally, the role of the subsistence function has decreased and production for the market has become increasingly significant. The importance of non-economic values of farming associated with the family-based farm has also been eroded as farms became closely tied into the market system and farmers and their families adopted some of the values associated with capitalistic production. In this context, changes in actual and perceived monetary returns to farming and farm employment, and in the life-styles that can be supported by them compared to non-farming occupations, become important indicators of the utility of farming as an occupation.

More recently, other values, mainly non-economic in character, have become more significant for some people involved in farming, e.g. the pleasure associated with working directly with the land and/or with caring for the environment. Whereas declines in non-economic values associated with farming have been linked with the rise of industrial society, their re-emergence appears to be strongly linked to the development of post-industrial society. And if the apogee of post-industrial society is to be found in the development of the regional city and its dispersed urban field or City's Countryside (Bryant et al., 1982), then we can expect to find a greater development of such forms of agricultural production in the City's Countryside than elsewhere.

THE HUMAN RESOURCE BASE IN AGRICULTURE IN THE CITY'S COUNTRYSIDE. Farm labour and management in the City's Countryside are subjected to three interrelated stimuli: employment, life-style and values. Employment considerations have been the most studied, although usually not explicitly in the geographic context of the City's Countryside. The development of non-farm employment opportunities can create incentives for farm labour and management to withdraw from agriculture. While the presence of non-economic values for involvement in farm production will introduce some stability into the pool of farm labour and management remaining in agriculture, it is obvious that playing down these values will encourage farm labour and management to withdraw from agriculture. So, there are both pushes (e.g. level of returns on farming, which are influenced both by revenue – and therefore product prices – and by costs of farm inputs) and pulls (e.g. incomes to be earned outside agriculture, new life-styles that can be attained).

It is these sorts of considerations which have led to the general decline of the human resource base in agricultural production throughout the Western World, and, together with the rise of first the secondary and then the tertiary and quaternary economic sectors, to the declining importance of agriculture as an employer. Hodge and Whitby (1981) note, for example, that between 1965 and 1977, the proportion of the active labour force employed in agriculture declined from 5.6 per cent to 2.7 per cent in North America, from 24.6 per cent to 11.4 per cent in Western Europe, from 9.6 per cent to 6.2 per cent in Australia and from 13.2 per cent to 9.7 per cent in New Zealand.

The geography of changes in the human resource base employed in agriculture is much more complex however. This reflects the fact that the stimuli that give rise to changes in the agricultural labour force are not uniform geographically. First, the structure and level of prosperity of agricultural production varies regionally and sectorally, so that the 'pushes' associated with movement out of farming will vary. This is important not only for understanding variations in response in terms of the economic factors (e.g. income potential) but also in terms of life-style considerations. For example, the work involved in dairy farming is much more demanding and continuous than cash cropping. Therefore we expect that dairying will be more sensitive to the spread of income opportunities and associated life-styles. It is not surprising then that various observers have noted the relative decline in dairying in the City's Countryside, despite the advantages of close proximity to a market (Berry, 1979; Johnston, 1989). In other sectors, there are still opportunities for agricultural intensification, as our discussion on enterprise structure in the City's Countryside indicated earlier in this chapter. Where these exist, they will help stabilise the total agricultural human resource base.

The other aspect of the debate on the reduction in agricultural employment concerns the presence and distribution of alternative employment opportunities in the non-farm sectors. Smit's analysis (1979) for Canada for the period 1946 to 1973 revealed a positive relationship between withdrawal from farming and favourable non-farm employment opportunities and wages for farmers, unpaid family labour and hired labour. This is a general phenomenon that is not restricted to the City's Countryside, and reflects the general commodification of farm labour accompanying industrialisation and urbanisation.

However, the distribution of alternative opportunities is most definitely uneven geographically and it is reasonable to assume that the opportunity costs of remaining in agricultural employment are considerably higher where there is a close relationship between the farming areas and the urban-industrial complex. For example, the precipitous decreases in farm population in the Paris basin during the early stages of urbanisation and industrialisation in France in the nineteenth century have been linked directly to the increasing levels of non-farm job opportunities there, despite the fact that the agricultural incomes and wages measured on an absolute scale in that region were amongst the highest in the country (Pautard, 1965). We shall discuss the effects of this phenomenon in terms of farm structure and the land in the concluding section to this chapter. Interestingly enough, the distribution of alternative job opportunities also has an impact on the human resource base in farming which works in the opposite direction, and this involves the development of various combinations of farming with a non-farm activity.

PART-TIME FARMING, HOBBY FARMING AND ALTERNATIVE AGRICULTURE. Part-time farming or multiple job holding including farming has deep roots in Western agriculture, and is a natural development in temperate latitudes, given the seasonal variation of demands upon the farm labour supply. By contrast, the emergence of specialised, agricultural areas dominated by full-time farmers is a relatively modern phenomenon, related to the development of an industrial society. However, with increasing urbanisation of society, and the development of post-industrial society, part-time farming has become a significant component of the agricultural structure again (Mage, 1982; Tweeten, 1983). For instance, in France 57 per cent of farmers were full-time in 1970 but in 1988, this had declined to 50.2 per cent (and only 17.5 per cent of spouses and 30.3 per cent of other farm family members) (Houée, 1990). Part-time farming, or pluri-activity, had almost become respectable again!

As with change in the agricultural population generally, it is useful to think of part-time farming as being a response to a series of 'pushes' and/or 'pulls'. On the push side, farm income opportunities are important, and it is not surprising to find high levels of participation in off-farm work by farmers in some of the poorer, more marginal environments for agriculture, e.g. parts of the Maritime provinces in Canada (Wong, 1983).

On the pull side, however, the opportunities for involvement are more numerous and often more substantial in areas directly in contact with the urban–industrial complex. This is why part-time farming is so often thought of as an urban region phenomenon. Wong's (1983) analysis of agriculture in Canada's metropolitan regions revealed a much higher level of off-farm work in terms of days of off-farm work per farm reporting off-farm work than in the hinterland areas. Similarly, the greater incidence of part-time farming in a recent study of an area south-west of Toronto compared with the Montreal south bank region can be partially explained by the higher level of off-farm opportunities in the Toronto region (Marois et al., 1991). These observations are all consistent with the contention that agriculture in the City's Countryside comprises several different forms of socio-economic organisation of production, with significant numbers of full-time, often specialised farms coexisting with different degrees and types of part-time farming.

In relation to this, part-time farming does not appear to represent only a transitional form of operation, either out of or into farming. This underscores the notion that some of the socio-economic forms of production included in part-time farming are significantly different from the full-time operations. Mage (1982), for instance, suggests that part-time farming is not a primary means of exiting from farming, and supports this with evidence from Ontario that shows that less than 5 per cent of the full-time farmers in 1966 who had left agriculture by 1976 actually passed through a 'part-time' status in 1971. On the other hand, a larger number of part-time farmers, at least in Ontario's case, become full-time farmers (20 per cent of Ontario's part-time farmers in 1966 were full-time by 1976), using the part-time activity as a means of easing into farming. This still leaves a significant portion of part-time farmers who remain in that position.

It is these permanent part-time farmers who emphasise the existence of different forms of socio-economic organisation of production. However, even this group is far from homogeneous (van den Berg, 1991). It includes the urban/industrial worker who develops a farming enterprise to supplement family income or to provide an additional activity and source of income during retirement. The worker–peasant smallholdings close to many West European cities provide good examples, and their development has been accentuated in some instances by generous ('early') retirement packages in industry or in public service (e.g. the smallholdings that developed in the Chevreuse valley south of Paris, many of whose operators were employees of the regional public transport system) (Phlipponneau, 1956; Tricart, 1951). More generally, then, one of the types of permanent part-time farmer is someone who takes on farming, or who takes on an off-farm job, to provide a permanent source of supplementary income for the family.

The permanent part-time farmer also includes the hobby farmer, whose primary motivation for farming is not income related, but rather the pleasure to be gained from engaging in farming. Not unnaturally, these sorts of part-time farmers are heavily concentrated in the City's Countryside because of the strong link with the primary source of family income which is characteristically employment in the non-farm sectors (Archer, 1978; Troughton, 1976a, 1976b; Wagner, 1975).

Part-time and hobby farming is frequently associated with a lower level of labour input. Moran (1988) in his New Zealand study showed how dairying in his sample was associated with a lower incidence of off-farm work. The relationship is potentially a two-way one, of course, with off-farm work leaving less time for labour-intensive enterprises and more labour-intensive enterprises using up a greater proportion of the on-farm labour supply. When the link between off-farm work and lower labour intensity is due to a lower level of intensity of production than on full-time farms with the same enterprises, rather than a mix of enterprises requiring lower labour inputs, the part-time and hobby farming phenomenon can be seen as contributing to the movement towards 'alternative agriculture'. This is the development of farming practices which contribute to sustainable agricultural production systems by using technologies and practices less damaging to the long-term potential of the soil. This phenomenon is not, of course, confined to part-time or hobby farming.

Finally, it is important to note that participation in off-farm work is not confined to the farmer alone, even though that is how 'part-time' activity has usually been defined (i.e. on the basis of the minimum amount of time spent on and/or income derived from the off-farm activity of the farmer). The farmer's spouse or children also may participate in off-farm work, and therefore contribute to family income (Bollman, 1988). The extent to which this can develop is partly dependent upon the availability of off-farm opportunities, and this is the basic reason for the positive association that has often been reported between level of rural farm family income and non-farm job opportunities (Ruttan, 1955; Sisler, 1959). So, once more, we can expect important ties to develop between the farm family and the urban complex, ties which serve to challenge stereotypical images of what constitutes the 'traditional' farm family in the City's Countryside – or elsewhere for that matter.

We return to consider some of the implications of the part-time farming phenomenon at the end of this chapter in some summary comments on responses from the agricultural sector to the various pressures and stimuli associated with the various market systems of exchange.

Capital in agriculture

FARM TECHNOLOGY AND AGRICULTURE IN THE CITY'S COUNTRYSIDE. Farm technology has evolved rapidly and continually over the last century and a half, and has been intimately connected to the development of industrial society. Indeed, the urban-industrial complex has often been seen as the source of many of the developments of agricultural technology (Boserup, 1965; Schultz, 1953), particularly those involving mechanisation and motorisation, but also of the whole host of developments in fertilisers, herbicides and pesticides, and, more recently, in biotechnology research.

There are two broad types of agricultural technology from the perspective of their links with other components of farm structure, viz. labour-reducing technology and labour-intensive technology (Pautard, 1965). Labour-reducing technology refers to the various types of mechanisation and motorisation that have been developed, leading principally to capital substitution for agricultural labour. In many cases, production levels remain stable, except where they lead to increased ability to cut losses from more timely field operations and improved storage operations. Characteristically, these labour-reducing technologies involve 'hard' technology, in the form of relatively 'lumpy' equipment. What this means is that scale economies result from their utilisation, and, therefore, from an economic perspective they are more easily integrated into larger farms than smaller farms. When they are integrated into smaller farms, this can lead to overcapitalisation and inefficiencies in farm production and in the allocation of farm resources within the production system.

Labour-intensive technology on its own refers to those developments in agricultural practices that lead to an increase in the amount of labour required. In the past one hundred and fifty years, these have also been associated with increases in levels of production. We are referring here to the increased use of fertilisers, herbicides and pesticides, as well as a wide range of 'soft' technologies in farm management. In the latter case, improved farm management practices include better accounts management, better adjustment of levels of application of fertilisers and so forth to soil conditions, improved pasture management and the introduction of various conservation tillage practices. In some cases, the practice clearly increases the level of labour input per unit area; in others, there may be a reduction in the physical labour input, but a substantial increase in the management component, for example, in the introduction of minimum or zero tillage practices. An important characteristic of the labour-intensive technologies is that they tend to require lower farm size thresholds to be economically and technically viable; they are therefore more adaptable to a wider range of farm sizes.

Often, labour-reducing and labour-intensive technologies occur together so it is not always obvious what the net effect will be. The decrease in the aggregate supply of farm labour is, however, quite generalised, suggesting the dominance of labour-reducing technologies. An interesting phenomenon that has been observed to accompany the earlier increases in farm size has been an increase in the hired labour force even though the aggregate farm labour supply decreases (Smit, 1979); such changes are consistent with the general move towards an industrialised and capitalistic form of farming.

However, if stabilisation of the land base occurs, continued mechanisation and motorisation results in a decrease in the hired labour component of such farms, and concomitantly increasing their reliance upon family labour. In this case, while part of the farm structure appears to be moving back towards a more family farm type of structure, it would be naïve to see in this the demise of the industrial mode of farming. This phenomenon has been reported in the United Kingdom recently (Marsden et al., 1989), and it also appears such farmers tend to have more frequent recourse to contract services for the specialised skills often required by their wider use of a greater variety of both hard and soft technologies (Errington, 1986).

Furthermore, some technologies are enterprise specific while others are more generally applicable to a wider range of enterprise. The tractor is the perfect example of a technology that can be integrated into many types of enterprise, and, given the range of machine size available, used on farms of varying sizes too. In contrast, the combine harvester and silo storage installations are much more enterprise specific. To the extent that there are broad regional differences between different Cities' Countrysides in terms of agricultural specialisations, then we can expect the rate of diffusion of some types of farm technologies to vary regionally too, as well as their associated labour effects.

Thus, because of the links between certain elements of farm structure, notably farm size in terms of land base and business size, and the integration of farm technology into the agricultural system, any change that influences those elements of farm structure will also influence the adoption of technological change. Thus, where local and regional conditions favour labour withdrawal from agriculture and farm consolidation, conditions can be created for the successful integration of capital-intensive and labour-reducing technology into the farm. In areas such as the Paris basin which experienced substantial labour withdrawal from farming, including both hired labour and farmers, during the nineteenth and twentieth centuries, the levels of farm consolidation and farm mechanisation are amongst the highest in France (Brunet, 1960; Bryant, 1973b).

While the explanation is more complex than simply differential labour withdrawal, it is interesting to note the pattern of change in some of the key elements of farm technology diffusion in Canadian metropolitan regions compared with their hinterland regions (see Figure 3.3) (Wong, 1983). Tractors clearly diffuse rapidly across the country: although there is not much difference between metropolitan and hinterland regions, metropolitan regions exhibit the highest levels of initial adoption. The presence of trucks and cars on census farms shows a similar pattern; these two items also reflect the concentration of hobby farms in metropolitan regions. Combine harvesters, on the other hand, diffuse at similar rates to start with and then continue to develop, relatively speaking, in the hinterland regions while their level of diffusion levels off much earlier in the metropolitan regions, reflecting the greater variety of farm types in the metropolitan regions as well as the important Canadian grain areas in the Prairies which fall beyond the metropolitan regions. Similarly, Gregor's (1981) analysis of capitalisation in Southern California stresses that proximity to urban areas, more so than good farmland, has a strong positive link with capitalisation in agriculture because of the emphasis in such areas on cropping, intensive production and managerial efficiency.

The integration of capital-intensive agricultural technology has had a number of other effects on farm structure. As the industrialised model of

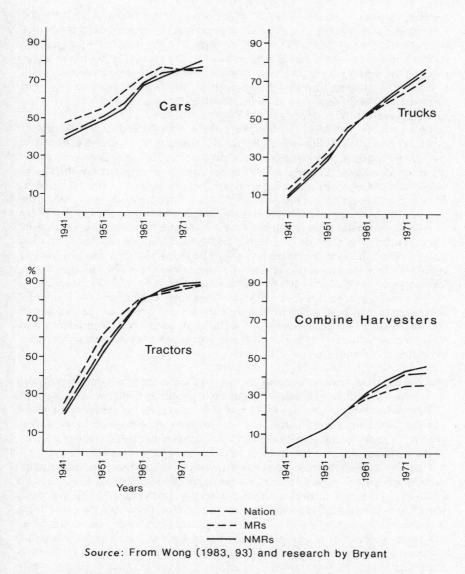

Figure 3.3 Diffusion of selected items of farm machinery in the metropolitan (MR) and non-metropolitan regions (NMR) or Canada, 1941 to 1976

farm production has become more and more prevalent (Troughton, 1982a), the use of non-farm inputs in agricultural production has increased substantially. Not only has this been linked to changes in the volume and nature of the labour input in farming, it has also increased the role of capital in the farm structure (Houée, 1990; Marsden et al., 1986a; Munton et al., 1988). Capitalistic farm production depends more on credit, and needs it, to acquire the expensive pieces of machinery characteristic of the modern industrialised farm. The net result of this is a farm production system in which credit is essential to the farm's survival, and in which farmers have become caught up in a technological treadmill, seeking improvements in productivity only to find themselves increasingly at the mercy of external forces (e.g. as reflected in interest rates) (Munton et al., 1988).

This process provides yet another avenue by which external blocks of capital penetrate farm business. The industrialised farm production model, a capitalistic mode of production, is thus an increasingly dependent one, which must watch almost helplessly as external forces modify the conditions in which it accesses credit (e.g. interest rates, credit policies) and the prices it pays for its inputs (e.g. energy costs, costs of machinery). Ironically, this penetration by external capital and the transfer of an important control point outside the farm has been encouraged by government intervention in country after country (e.g. Canada (Caldwell, 1988), the United Kingdom (Marsden et al., 1986b) and France (Houée, 1990)). The industrialised farm production model is therefore a relatively fragile production system. In the City's Countryside, where conditions have favoured and even encouraged its development in some sectors, we should not forget this fragility. It may be that some of our 'best' farm models are also some of our most vulnerable and the implications of this for land use stability in the City's Countryside and how we view different modes of socio-economic organisation of farm production are enormous.

THE AGRICULTURAL SERVICE NETWORK IN THE CITY'S COUNTRYSIDE. Because of the increasing dependence of industrialised farm production on external supplies of non-farm inputs, the network and infrastructure for delivery of these supplies form an integral component of the broader agricultural production system. It includes the distribution and accessibility of credit facilities, as well as the distribution of agricultural supply points and service facilities.

Changes in the networks of agricultural services can therefore potentially influence costs of access to these services for the farmer. Major changes have occurred in the agricultural service network generally, largely related to the complex processes of rationalisation. These have been characterised by a decline in the number of service outlets (e.g. farm machinery sales outlets) as the scale of individual outlets has increased in a market that has become increasingly stagnant. Changes in demand from the agricultural sector also vary regionally and are related to other changes in farm structure; one indication of this is the differential decline in the number of farms regionally. This can have an impact on demand, but the relationship is a complex one. Decreases in the number of farms can be compensated for by increases in farm size and needs of the individual farm operation. However, decreases in the level of agricultural production are likely to have a more direct impact on demand for agricultural services and number of outlets.

The most obvious type of decline in regional production comes from decreases in the area of agriculturally productive land used when they outweigh any increases due to intensification. As with retail services, agricultural service outlets require a minimum level of demand for viable operation. As land is removed from agricultural production, demand may decline below minimum threshold levels. This is turn may further undermine the viability of remaining farmland and farms by increasing their access costs to more distant service outlets. This is not likely to be significant for acquisition of the larger, higher-valued pieces of machinery and equipment, but is likely to become more than a nuisance for other types of needs, e.g. machinery repairs. Beyond the City's Countryside, situations exist where the decrease in farm numbers and agricultural labour has been associated with wholesale farmland abandonment, undermining the agricultural service network and the continued viability of farm communities (Berry et al., 1976; Crickmer, 1976). Similar concern has been expressed in parts of the City's Countryside due to agricultural land conversion, but no systematic evidence of this relationship exists. It might be expected to be relatively localised, given the fact that active conversion of land is itself quite localised.

No discussion of agricultural service networks would be complete without some mention of agricultural service centres generally and the non-farm needs of the farm family. Retailing has undergone significant rationalisation, related both to scale considerations and to the increasing mobility of the consumer. The result has been a decline in the number of outlets. In agricultural areas where this has been combined with declining farm population and increasing farm size, the market for many agricultural service centres has declined below necessary thresholds, leading to a decrease in the variety of services available and in the number of service centres (Hodge, 1974). However, in the City's Countryside, where there has been an influx of exurbanites, market sizes have been maintained or have even surpassed minimum thresholds, as exurbanites inject some of their spending into the local rural area (McRae, 1977). Under such circumstances, 'rural' service centres have stabilised and even blossomed. For example, Coppack (1985) describes for the Toronto-centred region a situation of declining rural service centres from 1941 to 1961, followed by what he calls a 'renaissance' of many rural centres after 1961 as the exurban movement picked up speed and more than compensated for the decline in the agriculturally-supported population.

Land

Land is the input in agricultural production which has received the most attention: (1) in the context of the City's Countryside and the phenomenon of agricultural land conversion (see Chapter 2); and (2) in the context of the impacts of modern agricultural technology on the soil (Brklacich, 1989; Coote et al., 1981). In this section, we consider both the value of agricultural land, its links with other inputs (intensity of land use) and the way farmland is incorporated into the farm (land tenure).

THE VALUE OF AGRICULTURAL LAND IN AGRICULTURAL USE. It is important to distinguish between the value of agricultural land for agriculture from the market value of land, as well as from the non-market values of agricultural land. In terms of the value of land for farming, we must recognise the link between the value of land in a particular use and the value or net returns that can be made from that particular use or farming system (Barkema, 1987; Hamill, 1969). The future stream of expected agricultural returns can be discounted and capitalised to yield a current value of land in use. The greater the expected income, the greater the value of the land in that agricultural use or system. The demand for agricultural land is therefore a derived demand, and values for land in agricultural use can be expected to reflect changes in agricultural commodity prices. This is particularly noticeable where there are few potential alternative non-farm uses for the land, i.e. when the agricultural use value approximates the market value of the land.

Given a particular set of commodity prices, more 'productive' land will command higher values. This is the basis of the initial Thunian formulation of agricultural land use patterns around a central market-place (Chisholm, 1962; Hall, 1966). Thus, land which yields a higher net return (i.e. is more 'productive' for whatever reason) will command higher prices. The simplest formulation of Thunen's analysis focuses on produce transportation costs which gives the classic net return or 'potential rent' schedule shown in Figure 3.4(A) for a particular farming system or use (Hall, 1966). Considering additional agricultural uses gives different potential rent schedules. These have different slopes because transport costs to the central market-place have a different impact on different products depending upon the bulk, weight and/or fragility of the particular product. They also have different intercepts with the vertical axis because in the absence of transport costs, different uses will generate different levels of net return depending upon the specific interplay of product prices, costs of production and yields per unit area.

In the initial formulation with the assumed homogeneous land resource base and closed regional system of agricultural production and exchange (the 'Isolated State'), competition leads to the potential 'rent' being bid away in actual rental payments (the basis for the land values) for the land. The result is the familiar concentric land use pattern (Figure 3.4(B)), in which the use generating the highest net return in any particular location captures access to the land resource. Relaxing the assumption of the homogeneous land resource base gives rise to a mosaic of land uses. Relaxing the assumption of a closed regional system of production and exchange brings more distant producing regions into the potential supply area for the urban market if their production costs are lower and outweigh the additional transport costs.

Even considering land values for agricultural uses alone is more complex than this discussion suggests. Farmers may be prepared to pay a premium for land with more potential alternative uses because of the greater flexibility this might provide them. On the other hand, the real differences in the land potential are usually not reflected in the same degree of differentials in the values paid for land for farming purposes.

Farmers often pay a premium for smaller parcels of land (see, for example, Bryant (1982)) either because they round out a particular field or raise the

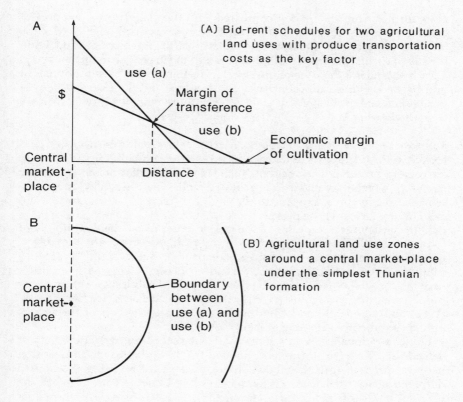

Figure 3.4 The Thunian formulation of agricultural land use location

total productive land base of the farm to a level that permits more effective use of machinery, equipment and installations. This point emphasises the difficulty of separating the 'pure' value of the land from the contributions of other factors of production or inputs to the productivity of the land. Examples of such inputs which are fundamental to the productivity of the land include tile drainage, on-farm infrastructure such as fences and lanes and, of course, all the labour that has been invested in them. It is no wonder that the data produced by the Valuation Department of New Zealand which give 'unimproved' values (the market value of the land less improvements) are so hard to replicate elsewhere (Moran, 1978).

Intensifying the use of agricultural land is a rational (economic) strategy when net returns increase as a result. Under the initial Thunian framework (Hall, 1966), each level of intensity of production of a given product can be seen as a separate farming 'system'. More intensive systems yield a higher level of net return nearer the market (assuming, of course, that production is rational and does not occur in situations of negative returns to increasing capital and labour inputs). However, they are subject to higher transport costs further from the market compared with the yield from a unit area of land under a less intensive system for the same product. This is why the intensity component of the model predicts that the intensity of production for any particular enterprise is higher closer to the market than further away. Clearly, under these circumstances, relaxing the assumptions of homogeneous physical productivity relationships and the closed system of production and exchange will yield an even more complex mosaic than we suggested earlier.

2 — **NON-FARM VALUES OF AGRICULTURAL LAND.** Market values exist in agricultural land for functions it supports other than its productive agricultural value. There are other 'production' functions such as the use of mineral aggregate resources which may underlie agricultural land. There are 'play' functions, which also add a market value to the land, e.g. recreational enterprises and even hobby farms whose prospective owners may be prepared to pay more than the agricultural use value of the land in order to pursue their hobby of farming. Finally, and of generally greater importance, are the 'place' functions of land whereby the land is in demand because it simply provides a support for residential, commercial or industrial uses. Hence, characteristics such as accessibility to employment and shopping facilities, as well as site or property specific characteristics of the land influence values (see Bryant et al. (1982) and Goodchild and Munton (1985) for a discussion of the factors influencing the market value of land).

Under a rational economic framework, market values reflect the highest valued use. We have already encountered this in outlining the competition between different agricultural uses. In the context of land being allocated to different agricultural uses, changes in relative prices – and therefore the relative levels of net returns for different enterprises – lead neither automatically nor immediately to changes in agricultural land use. This is partly because of inertia that builds up in the agricultural system, either for technical reasons such as the relative fixity of certain farm assets or for socio-cultural reasons, and partly because different enterprises face different degrees of uncertainty about prices and yields. Similarly with respect to non-farm values of the land compared with agricultural values. Generally, non-farm or urban uses of the land give rise to higher market values. There are exceptions, however, e.g. the short-lived period when land for kiwi fruit production in New Zealand in the 1970s commanded much higher values than land destined for suburban development (Jackson, 1985), but these situations are rare.

The extent of the demand for land for non-farm purposes in the City's Countryside is influenced by the growth potential for economic development in the settlements and their peripheries in the regional city. There are many different submarkets for land for non-farm purposes which overlap with each other. Different sets of factors influence the demand for land in these different submarkets. Accessibility considerations of various types are often important, but so too are other characteristics in some of the submarkets. Lot size, the existence of rolling topography, woodlots and proximity to water bodies and streams or rivers, for instance, are important factors that help shape demand for residential properties in the City's Countryside (Diemer, 1974; Gibson, 1977; Moncriff and Phillips, 1972; Rajotte, 1973). The interplay of the geography of the supply of the lots with particular characteristics can also yield complex and scattered patterns of non-farm development.

Non-farm development is, therefore, far from a simple, geographic process. Even the accretionary patterns of suburban development – residential subdivisions or estates, industrial and business parks, and shopping centres or malls – do not proceed along simple geographic lines. One set of factors that can influence these patterns is related to the agricultural structure itself. For instance, agricultural structures do sometimes present resistance to development pressures, either where the agricultural structure has remained prosperous or where other factors have helped stabilise farming, e.g. cultural or personal attachment to a particular area. At the least, such factors introduce short-term rigidities into land development, and at the other extreme, they can influence permanently the future form of the urban landscape or even be maintained as some form of open space enclave.

The observation of actual market values for land which are greater than farmland values has often been suggested as an indicator that farmers will see their land as being ripe for conversion (Rodd, 1974, 1976). While this may be a reasonable expectation close to the urban edge, little evidence exists for this in more distant locations. Objective considerations of the absolute area of land for, say, country residential development in relation to the total area of agricultural land emphasises how unlikely such conversion is for most land.

However, the observation of real estate transactions of farmland with values greater than the farmland value may be responsible for prompting an upwards evaluation of the likelihood of development on the part of the farmland owners. Where this situation exists, expectations are raised across a much broader geographic area than is ever likely to be involved in actual land use conversion. It is this argument and the accompanying thesis regarding the uncertainty of an agricultural future (or should we say an increased expectation of development?) which has led many observers to conclude that the negative impacts of potential urban or non-farm development on agricultural practices can be more damaging to farming productivity as a whole than the actual removal of land from agricultual production by land use conversion (Berry et al., 1976; Johnston and Smit, 1985; Krueger, 1978).

The actual removal of land from agricultural production in the course of land development and non-farm development generally has been referred to elsewhere as a direct impact of urban forces on agriculture (Bryant and Russwurm, 1979; Pacione, 1984). Other impacts related to the adjacency of non-farm development which influence the viability and productivity of the ongoing agricultural structure but which do not lead immediately to land use conversion have been termed indirect impacts. We now consider how these indirect impacts may influence the viability and productivity of the farm system. We must emphasise, however, that not all of the impacts are unambiguously negative.

THE ANTICIPATION OF NON-FARM DEVELOPMENT AND ITS IMPACTS UPON FARM PRACTICES. Attempts to utilise the classic Thunian framework to handle the conversion of agricultural land to residential development (e.g. Muth, 1961) provided valuable insights into the role of the relative inelasticity of demand for housing versus the greater elasticity of demand for locally produced agricultural produce in accounting for the poor degree of resistance of agriculture to development pressures. However, in other respects, trying to understand conversion through such modelling efforts is unsatisfactory because of their unrealistic requirement for simultaneous and automatic adjustments in land use patterns to changes in the parameters such as relative prices of housing and agricultural produce. In reality, lags are common, and conversion, as alluded to above, is far from being a clean, efficient process.

Indeed, the onset of continual or increasing degrees of pressure for non-farm development, especially in terms of accretionary urban development, not only influences the market value of land because of the increased potential to support urban development (Rancich, 1970), but has also been held to decrease the value of the same land for agricultural purposes (Sinclair, 1967). Thus, it has been suggested that the shortened time-scale for agricultural investment planning, given the expectation of imminent development, can lead farmers to 'farm to quit', whereby they 'mine' the agricultural productivity – the combination of the natural fertility of the land and the accumulated positive impacts of the application of labour, management and capital to the land (Wibberley, 1959). Conversely, if planning intervention effectively restores 'normal' investment planning horizons, e.g. in 'agricultural reserves' or in Green Belts, then we might expect agricultural use values to rise (Boal, 1970). This raises the more general issue of the role of public intervention in creating increases in land values. Munton (1976), for example, noted that it was common in south-east England for properties with planning permission to reach over twenty times more than their agricultural use value.

The first systematic effort to incorporate the influence of the anticipation of urban development factor into explanations of changing agricultural land uses in the urban fringe was made by Sinclair (1967).

As the urbanized area is approached from a distance, the degree of anticipation of urbanization increases. As this happens, the ratio of urban to rural land values increases. Hence, although the absolute value of the land increases, the relative value for agricultural utilization decreases. (Sinclair, 1967, 78).

Sinclair argued that because the role of transportation costs of moving farm produce to the market was no longer of great importance, the economic rent of land for agricultural investment close to the expanding urban edge would be lower than further away from the expanding city (Figure 3.5(A)). In cases where the development of the land was expected imminently, the result could even be disinvestment in the land resource (along the lines suggested by Wibberley (1959)), possibly resulting in premature setting-aside of the land.

The basic argument was elaborated upon by Bryant (1974). Bryant argued

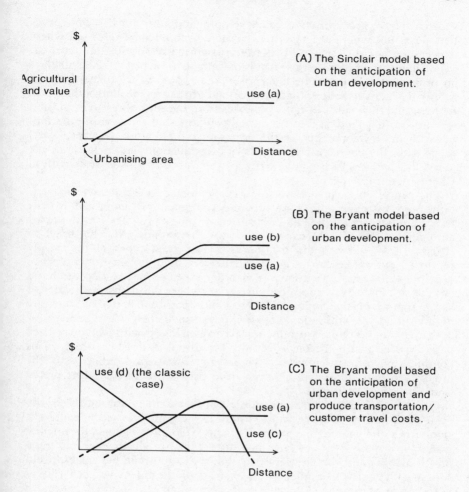

Figure 3.5 Agricultural land value curves around an urbanising area

that in environments characterised by rapid urban development, certain types of agricultural investment could indeed be down-valued by the shortened planning horizons there. This would influence those types of investment with a high level of fixity and long amortisation periods. Examples of investment that would be affected by such shortened planning horizons are new buildings and building maintenance, field drainage, new orchard plantations (see, e.g. Bryant, 1974) and grain silos (Berry, 1979). This yields value curves for agricultural land that are similar to Sinclair's, except that the differences in the angles of the upward-sloping portions of the curve are related to the degree to which the investment required for each particular use or farm system is affected by having its planning horizon shortened by expectations of urban development (Figure 3.5(B)). The curves flatten out not where expectation of urban development is zero, but where expectations are such that 'normal' planning horizons can still be anticipated.

Furthermore, if we recognise that proximity to the urban market still plays an important role for some enterprises, then some value curves can still follow the 'classic' pattern (no anticipation of urban expansion), others will first rise and then fall as distance increases from the edge of urban expansion (Figure 3.5(C)) and still others might follow Sinclair's configuration. Subsequent developments of the conceptual framework (Bryant, 1981) focused on the degree of homogeneity of evaluation and response of the individual farmers to the potential urban expansion. This is dealt with in Chapter 4 in more detail.

The process by which the value of land for agricultural investment is influenced by expectations of urban development as suggested by the frameworks just outlined is intuitively appealing. The evidence is ambiguous. This is to be expected because the framework as modified by Bryant (1974) allows both for increases in certain types of intensively produced enterprises in areas where there is high expectation of urban development as well as decreases in other types of intensively produced enterprises. Intensively produced enterprises where the intensity is related to labour inputs or capital inputs with a very short pay-back period or which are highly mobile are still rational possibilities in such areas. Conversely, new orchard plantations, dairy production and capital inputs with a long pay-back period such as tile drainage would be curtailed in such areas. This complication means that studies where general statistical indicators of the intensity of agricultural production have been used must be interpreted cautiously (e.g. Mattingley, 1972).

Other complications, apart from the fact that farmers do not necessarily evaluate and respond to the same stimuli in the same way (see Chapter 4), stem from the inertia that exists in many farming systems, the greater mobility of many types of farm investment than might be expected upon initial analysis and the temporal variability of the urban expansion itself. Moran (1979), for example, in an analysis of the intensive orchard and vineyard areas near to Auckland's western suburbs, observed continued normal functioning of these enterprises. This he attributed partly to the inertia deriving from the large amount of capital investment already in these enterprises. This tendency is reinforced when the actual loss of investment in farming is dwarfed by the very large capital gains to be made from sales of the land for urban development purposes. It is not without cause that some farmers look upon increasing farmland values in a very positive light (McRae, 1980, 1981; Sargent, 1970; Smit and Flaherty, 1981)! Furthermore, some investments are much more mobile than they appear at first sight. Examples include nursery production (especially when the trees and bushes are planted in pots), greenhouses and many types of irrigation system. Finally, urban growth pressures have not been consistent over time in many areas. For instance, in southern Ontario, growth slowed during the latter part of the 1970s only to pick up again after the recession of the early 1980s, and in the Paris region, France, the rapid growth of the 1960s was not continued through into the 1970s. Echoing this, in a Netherlands case study, Lucas and van Oort (1991) found in their survey of farmers having sold land between 1970 and 1985 in an area near Utrecht that most of the farmers had simply carried on with farm investments: the uncertainties of development were such as not to make it worthwhile 'disinvesting'.

(A) Fragmentation by new road construction

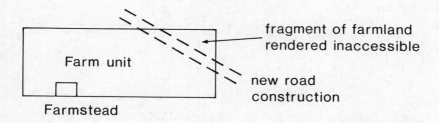

(B) Fragmentation by traffic and road improvement

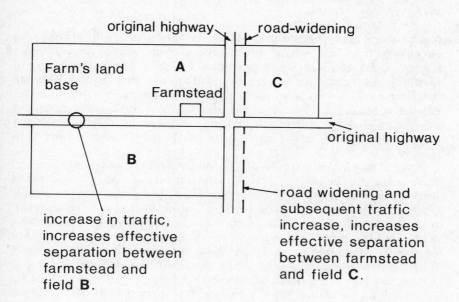

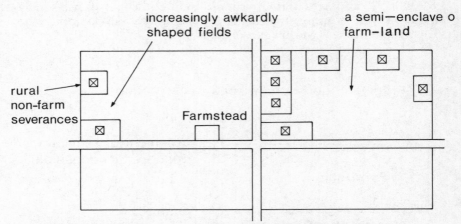

Figure 3.6 Different forms of farm fragmentation in a near-urban environment

It is not particularly easy to collect problem-specific data to test the anticipation hypothesis. Farm-level data under extensive control conditions would be required to sort out the influences of different types of farm enterprises, different agricultural environments and different degrees of urban development, not to mention the existence of various other impacts, some positive, some negative, that impinge upon agricultural investment in the City's Countryside.

OTHER IMPACTS UPON THE VALUE OF LAND FOR AGRICULTURE. Other influences emanating from the nearby urban environment include several other negative indirect impacts upon agricultural structure. These can combine to reinforce the effects noted above regarding the anticipation of urban development. They are not necessarily related simply to distance from the expanding urban edge, and can be linked to scattered non-farm development as well as accretionary urban development. They can, however, be expected to be more intense in the inner parts of the City's Countryside than elsewhere.

Fragmentation of the farm landscape and the farm unit can occur in several ways near a city (Figure 3.6). First, the convergence of road networks in the urban fringe of many major cities increases the probability of simple fragmentation by new road construction. Parcels of land become separated from the farmstead making access more and more difficult. In some cases access becomes impossible, even after considering the transfer of the parcel to another farm, and land is abandoned. Where this occurs, it is not uncommon for the public agency acquiring the properties for the road construction to acquire this type of separated parcel as well.

Second, fragmentation also can occur with an increase in road traffic or the improvement of existing road surfaces which may encourage an increase in traffic. The result can be the same as if a new road were built: increased time taken to move farm machinery from one part of the farm to another and a greater risk of accidents with farm machinery. In some urban fringe locations, urban expansion has engulfed farmsteads and buildings of operating farms; and farming in the face of heavy traffic, traffic signals and one-way street systems requires particularly tenacious farmers or special adaptations on the part of the farmers (e.g. relocating the centre of field operations for the farm), or the municipality (e.g. one-way street systems except for farm machinery). Both these types of adaptations were observed by Bryant in fieldwork west of Paris in the early 1980s.

Finally, fragmentation can occur through scattered non-farm development which nibbles away at the edges of fields. In the absence of land use planning of any significance, this has led to the creation of quasi-enclaves of farmland (Figure 3.6) with access rendered more difficult as farm machinery increased in size. In our fieldwork around Toronto, many farmers identified traffic problems and congestion as the most serious problem they faced in farming (Johnston, 1989; Johnston and Bryant, 1987).

We must be careful, however, not to interpret farm fragmentation always as negative. Frequently, as farms expand their land base, it becomes more fragmented (Bryant, 1973a; Edwards, 1978). While there are costs involved in such fragmentation, they are costs incurred in order to reap an even larger benefit. For example, Bryant's survey (1973a) of a sample of fruit farmers in the northern suburbs of the Paris conurbation demonstrated that for a group of expanding orchardists whose farmsteads and initial land base were embedded in the urban area, the solution was to acquire extra land fifteen kilometres (ten miles) or so away. The net result was a marked shift in the operational centre of gravity of the farms to more modern orchards, permitting more efficient use of machinery.

The incidence of theft of agricultural produce and equipment, and vandalism of machinery and installations, has also been claimed to be more intense in the City's Countryside (Toner, 1979). Certainly, most farm surveys in the urban fringe make note of this as a negative impact identified by farmers. While incidents are real, it is not clear how significant they really are on their own for the financial viability of the farm unit (UK Ministry of Fisheries and Food, 1977). Incidents of theft and vandalism are an extreme form of incompatibility between land uses, although it is far from clear that it is the adjacency of farm and non-farm land uses that gives rise to the impact.

Other forms of incompatibility frequently noted relate to complaints by members of the non-farm community about farm odours and dust and noise from agricultural field operations (Coughlin, 1979; Kohn, 1990; Punter, 1976; Sargent, 1970). From the evidence, this appears not to be much more than a nuisance factor, except where the situation is perceived to be so significant that measures are taken in land use planning and management to separate farm from non-farm uses and limit certain types of agricultural development in proximity to non-farm development (see Chapter 5).

It is difficult to obtain convincing evidence on the seriousness of these various negative impacts on farming. Taken individually, they do not appear serious. In two studies (Blair, 1980; Johnston, 1989), urban fringe farmers were asked not only whether they had experienced any 'urban-based' problems, but also what their implications were for the structure and operation of the farm. While comments were made about the existence of such problems, including theft and vandalism, by many of the farmers interviewed (42 per cent of Johnston's sample reported some sort of 'urban-based' problem), few issues appeared to have caused the farmers to make significant changes to their operations.

Even when farmers' responses to exposure to a whole range of 'urban-based' problems have been analysed together, Bryant (1981) found that while there was a strong relationship between the intensity of exposure to urban-based problems and the level of urban development pressures, there was no link between the complexes of problems identified and changes that farmers had made on their farms. On the other hand, there are unquestionably areas where the combination of negative impacts of all types has led to significant problems for ongoing farming (Bryant, 1984a). It is likely that in the studies noted above, such areas are relatively localised and their presence has been masked by inclusion in broader geographic zones. In addition, studies of farming in the urban fringe characteristically focus their attention on the continuing farms, and therefore ignore the fact that there are areas in the inner urban fringe of major metropolitan areas where the impacts of these factors in combination has been so large that the land has simply been abandoned.

Yet another potentially negative impact frequently alluded to in research on agriculture in the urban fringe is the effect of non-farm development on property taxes paid by farmers (see, for example, Hennigh (1978) and Krueger (1957)). This impact has been the subject of much attention in North America where local property taxes often form a significant portion of local municipal revenue. In the absence of tax rebate programmes or preferential farm property assessment programmes and the like, the research conducted in the 1950s, 1960s and 1970s demonstrated that small, residential lots contributed less to municipal revenue in relation to the municipal services used than farm properties, despite having a higher per unit area assessed value than farmland (see, for example, Wellington Planning and Development Department, 1977).

This was so for two reasons. First, there was a tendency for the assessed value of farmland to reflect the inflationary effects created by non-farm development, even where assessment was supposed to be on the basis of 'in use' values. Second, in the City's Countryside the demands for municipal services are greater because of the larger proportion of non-farm development of all kinds, so that the tax bill is higher in these locations. It has been suggesstted that the higher service demands in urban fringe municipalities are partly due to the greater levels of services expected by exurbanites or non-farmers when they move into a rural area. However, the evidence on the real differences in service expectations between farmers and non-farmers is ambiguous (Smit and Flaherty, 1981). None the less, the perceived issue has led to many forms of public intervention to correct the presumed injustice of having the farm population bear a relatively heavier portion of municipal costs (see Chapter 5).

4 THE TENURE OF FARMLAND. We have already touched several times on one of the major systems of exchange, the land market, through which expectations and desires about the future shape of the use of land are frequently expressed. Land conversion at the urban edge and scattered rural non-farm residential development in the City's Countryside are the results in terms of land use of changing ownership characteristics. In the case of land development at the urban edge, the scale of development characteristically involves large blocks or assemblies of land today, and it is not uncommon for the land to remain in non-farm ownership and still be farmed long before conversion takes place (Chung, 1972; Putnam, 1962). In the case of scattered non-farm residential development, or the purchase of farmsteads for a weekend retreat, which are surplus to farming's needs, other parcels of land can move into non-farm ownership and remain so over the long term. This, together with what has been the almost inevitable increase in the market value of land, has given rise to concern over the impact of increasing non-farm ownership on farm structure.

Increasing non-farm ownership and the accompanying inflation of land values makes it more and more difficult for farmers to increase the size of their acreage by purchasing additional land. This is important in several countries such as Canada, New Zealand and Australia where there is a strong cultural bias for farmers to want to own the land they farm (see, for instance, the comments on the goals of Ontario farmers by the Special Committee on Farm Income (1969)). Farm size increase has been one of the most pervasive features of structural change in modern agriculture in the twentieth century, and has been linked to the accompanying processes of capital substitution for labour, mechanisation and motorisation of agriculture (Bryant, 1976; Grigg, 1984; Whatmore et al., 1986).

Increases in the physical land area farmed have been pursued in order to try to match the technical and economic requirements of the increasing scale of farm machinery during the middle half of the twentieth century. Therefore, any process that makes it more difficult for farmers to acquire additional land has been held to restrict agricultural 'progress'.

Evidence on farm size change from many metropolitan regions provides partial support for the existence of such a dampening effect. Bryant (1976), for instance, notes that in the Toronto region in Canada, the average physical land base per farm increased by only 3.9 per cent between 1961 and 1971, in the Vancouver region it remained stable, in the St Catharines area it increased by only 3.7 per cent and in the Montreal region it increased by only 14.3 per cent; this is in marked contrast to the 29.2 per increase noted for Canada as a whole. Clearly, there are other ways to expand farm business size, namely through intensification or adopting special marketing procedures. Furthermore, there are other factors that partially account for these farm size trends too, such as the frequently greater proportion of the smaller hobby and part-time farms in such regions. None the less, non-farm ownership of farmland and the increased land values do place a brake on farmers purchasing land to expand their acreage. A good example of this is provided by Reeds' early study (1969) of the Niagara Fruit Belt, Canada. He estimated that fruit farmers in the area at that time could afford a maximum of about

$5,600 per ha but that market values for the same land hovered around $15,000 per ha.

However, the operation of the very same system of exchange – the market in land for sale – in the City's Countryside has simultaneously helped fuel another system of exchange – the market in farmland for rent. Ownership of land is still favoured by many farmers to ensure access to productive land. However, ownership also brings additional costs since it not only gives access to the land for farming, but includes a number of other rights as well, including the right to sell land for other purposes and capitalise on increases in the capital value of the land (within the limitations, of course, of any regulations governing the future use of land or the operation of the land market). Renting the land, on the other hand, provides access to the farming resource at a lower cost than purchase would entail.

Under such circumstances, one might wonder why many farmers still prefer to own their land. One reason is not directly related to the farming operation at all, and that is the desire to own land in order to benefit from anticipated increases in its market value. In particularly inflationary urban fringe land markets, farmers are not likely to be able to afford to purchase land for this purpose on any scale. This no doubt accounts for the very low proportion of farmers acting as purchasers of land in the immediate urban fringes of Kitchener–Waterloo in south-west Ontario in the late 1960s (Bryant, 1982).

On the other hand, considerations of being able to realise capital gains in land values may well lead farmers to hold on to land they already own in the hope of even greater values. Such holding behaviour is at least partially responsible for some of the leap-frogging development patterns around many US cities in the 1950s and 1960s (Archer, 1973; Hushak and Bovard, 1975). A related reason for retaining ownership of land, or for purchasing additional land where land values permit, is as a hedge against inflation and to secure the value of a farmer's or farm family's capital (Rodd, 1976). Farmers often prefer to own their land because of the security of tenure that comes from owning the land input to the farm operation. Farm investment planning, as for any business development, necessitates taking a long-term view of amortisation and effective integration of lumpy investments into the farm operation. We have already noted that in urban fringe locations where development is imminent, the result can be to shorten farmers' planning horizons and reduce or curtail certain types of investment. With rented land, it has been argued that this tendency to disinvest will be exacerbated (Ironside, 1979; Thomson, 1981) where there is uncertainty concerning the future use of the land, and that increases in the rental of farmland pose problems for its long-term productivity because farmers will not care for rented land in the same way that they will for their own land.

A number of points can be made in relation to the farmland rental question. First, it is clear that the rental of farmland has become a major method by which farmers have expanded the physical land base of their operations. In many Western countries, rented land has accounted for an increasing proportion of the farmland worked and the number of farms with mix of rented and owned land has risen. This has been observed in Canada both generally (e.g. Bryant, 1976; Clemenson, 1985; Troughton, 1982b) and individually (e.g. Fielding, 1979; Greaves, 1984; Johnston, 1989), as well as in other countries (Hill, 1974).

Second, farmland rental provides a cheap way of expanding the acreage to take advantage of scale economies in production. This has occurred both in the urban fringe as well as beyond the City's Countryside (Bryant et al., 1984), although it has often been much higher in the urban fringe. Johnston's (1989) study of farmers in the western fringes of Toronto, for instance, showed extensive renting of farmland for cash crop farming in the Oakville area, a highly urbanised and rapidly urbanising zone. And Bryant (1976) in a statistical study of agricultural change in the 1960s in the counties of Waterloo and Wellington some 50 to 80 kilometres (30 to 50 miles) south-west of Toronto found that average farm size increased in line with the degree of farmland renting and with the growth in population (a crude indicator of increasing non-farm ownership of the land).

Third, while indicators of the increasing importance of rental farmland in many parts of the City's Countryside in many countries are obvious, the evidence of its impact on farm management practices is much less clear (Johnston and Smit, 1985). On the one hand, it is not surprising that in rapidly changing urban fringes rented land is often used for cash cropping (e.g. Johnston, 1989; Putnam, 1962). In some situations, land in non-farm ownership may not be farmed at all and is set aside (Thomson, 1981). It is important to point out that where setting aside of the land has occurred, some evidence suggests that its extent is quite limited (Dawson, 1982). On the other hand, where attempts have been made to probe whether rented land is treated any differently from owned land, no clear patterns emerge (e.g. Baker, 1989; Ironside, 1979). Baker (1989) studied the potential erodibility on an extensive sample of rented and owned farm plots in the County of Oxford in south-west Ontario. Although the level of erodibility on rented farmland was higher than on owned farmland, this was largely because the rented land was characterised by steeper slopes than the owned farmland, indicating the generally prosperous nature of farming in this area and the relative short supply of farmland for rent. And while Munton's (1983a) analysis of farm properties by tenure and their apparent management in London's Metropolitan Green Belt showed a link between 'short lets' and the lowest category of management, it is important to note that the categories of management quality utilised are related more to the physical appearance of the land than to the agricultural productivity of the land.

The lack of any clear link between rented and owned farmland in terms of maintenance of its productivity is especially interesting because most of the evidence cited above comes from North American studies. In North America, farmland renting is characteristically undertaken on short term agreements (Fielding, 1979; Greaves, 1984), often verbal agreements, with virtually no provision made for compensation to the tenant farmer for his investment if the land is sold for other purposes or transferred to another farmer. Under such circumstances, it might be expected that rented farmland would be treated much less carefully than owned land. While it is also possible to find cases where this is true – one farmer in Fielding's (1979) study stated that he preferred renting because he felt he could move on quickly if there were any signs of deteriorating productivity! – this does not appear to be the case generally. Some farmers in areas with a high potential for urban development may certainly mine their rented land, but they appear as likely to do the same with their owned land, and this regardless of whether they are in North America with the characteristic short-term leasing arrangement, or in, say, France, where farmland is rented under a complex legal arrangement which provides considerable security of tenure for the tenant farmer. Access to farmland on a very short-term basis does exist in France (land farmed *à titre précaire*) in which the legal barrage of protection for the tenant farmer does not apply, but such situations are characteristic of the extreme urban fringe and public or private land assemblies where development is imminent (Bryant, 1981).

It is relatively easy to identify differences in the uses to which rented and farmed land are devoted: it simply demands extensive farmer surveys. Explaining such differences is not, however, quite so easy because other factors may be associated with the rented parcels of land, which have nothing to do with the fact that they are rented. For instance, farms which have extended their acreage characteristically become more fragmented, as it is often difficult to acquire access to rented land adjacent to the 'home' farm (Sublett, 1975). The greater distances involved may therefore dictate different uses of rented land compared with owned land within the same farm unit, with the newly acquired more distant rented land being used for less labour intensive activities. Furthermore, there are additional difficulties involved in defining the effect of renting farmland on the management practices of farmers. Farm surveys are not in themselves adequate to obtain the level of detail required, and field-by-field surveys and analyses are indicated. Evidently, a great deal more detailed empirical work needs to be undertaken in this area.

IDLING AND ABANDONMENT OF FARMLAND. Land use conversion in the City's Countryside, and particularly in the immediate urban fringe is obviously not a smooth process with automatic adjustments in land use. The discussion of the possible effects of the anticipation of urban expansion at the urban edge has already introduced the possibility of declining farm investment in the face of shortened planning horizons. In extreme cases, land may be removed from agricultural production altogether. Whether this can be construed as setting aside or abandonment of the land is related partly to the time-scale involved, as noted earlier in Chapter 2. Where the land is simply removed from agricultural production as part of the transition from farmland to developed land, it has been set aside. Such transitional periods appear to be normal in land conversion and may be prolonged when, say, planning permission or the negotiation of financial arrangements for the development take longer than anticipated. Concentrations of farmland with signs of recent farm activity (e.g. corn stubble from the previous year's crop) adjacent to areas of rapid urban development can be interpreted in this light.

Abandonment of land takes place when it is removed from agricultural production but there is no likelihood of development in the short to medium term. This might occur when there is a combination of negative impacts from the nearby city which are so great that farming cannot be continued. At the same time, we have to recognise the existence of other factors that can lead to land abandonment for farming, viz. lack of farmers willing to take up the land because of the attractions of non-farm employment, poor field structure (e.g. the many narrow and small field strips found south of Montreal that have remained abandoned for many years) or low quality soils.

Studies that assumed that abandonment of farmland has not been important in the City's Countryside of metropolitan regions (e.g. Crerar's (1963) classic study of farmland 'loss' in Canada's metropolitan regions) were based on the belief that the agricultural quality of the land for farming in such regions was too high for abandonment to take place. However, subsequent investigations using a finer geographic filter and the same census statistics or air photo analysis of land use change have demonstrated that much of the removal of land from farming in some metropolitan regions is very much the abandonment of poor agricultural land (e.g. Bryant et al., 1981; Gierman, 1977). Ziemetz et al. (1976) also pointed out the economic and technological obsolescence of farmland as an important factor in the removal of land from farming in some of the fast-growing metropolitan regions in the United States, already noted in Chapter 2.

The evolving pattern

Farming in the City's Countryside participates in many different market systems of exchange: markets for agricultural produce, markets for farmland both for purchase and for rental, and markets for other agricultural inputs including labour, capital, machinery, fertilisers and seed. Our focus is on farming in the City's Countryside, a meso-scale phenomenon in the evolution of human settlements. However, as alluded to in Chapter 1, the market systems of exchange do not necessarily function on that scale. Some certainly appear more local than others, such as the market for fresh local agricultural produce, the market for land for residential development or the market for hired agricultural labour. Even these, however, have become increasingly influenced by forces external to the City's Countryside. Markets for fresh local agricultural produce can be influenced by competition from distant producing areas, local and regional residential land demand is inexorably tied to macro patterns of investment in economic activities, and the market for local hired agricultural labour can be influenced by general increases in standards of living. Farming therefore has to contend with a myriad of forces, many of which are external to the farm's immediate operating environment (Olmstead, 1970). Any interpretation of the resulting changes in farming, agricultural structures and agricultural communities has to keep this in mind. In the concluding part of this chapter, we offer some summary comments concerning the broad patterns that appear to be evolving in farming in the City's Countryside, before dealing more explicitly with the role and behaviour of the individual farmer and family in the next chapter.

STRUCTURAL PATTERNS. The different forces affecting agriculture in the City's Countryside, together with the potentially greater variety of methods of agricultural production noted in Chapter 1, could elicit many different structural responses from agriculture. Clearly, the industrialised model of farm production with its increasing scale of operation, either through increases in the acreage of individual farms or through intensification or both, and capital substitution for labour has found some conditions in the City's Countryside favourable to its development (Gregor, 1988; Lawrence, 1988). The removal of labour from farming has been influenced by the greater non-farm employment opportunities in urban areas, the scarcity of labour has encouraged mechanisation and specialisation at the farm level, and even in the immediate urban fringe, the frequently increased availability of farmland for rental has facilitated the consolidation of farms into larger and larger units.

While on a macro scale of analysis, broad regional specialisation in agriculture has occurred in many countries (e.g. Canada (Wong, 1983), the United Kingdom (Thomson, 1981) and New Zealand (Moran and Nason, 1982)), within the City's Countryside of some major cities at least, a greater variety of specialised farm units have developed in response to the greater variety and scale of demand in those areas (Gregor, 1988). These specialised types of farm are often very intensive operations with a relatively small land area, e.g. intensive poultry farming, feed lot operations, and market gardening.

At the same time, however, other sets of conditions in the City's Countryside have favoured the development of different socio-economic modes of production. While the family farm resembles more and more a capitalistic unit, it is still a fundamental component of much of the farming around cities. Its apogee can perhaps be seen in the intensive fruit and vegetable farms found outside cities such as Paris or in the Niagara Fruit Belt in southern Ontario where the units are intensive, but relatively modest in scale, where family labour is dominant and where the family works the farm and sells a significant portion of the produce in the suburban market-places or at roadside stalls.

Even more important in terms of the overall structural pattern that is evolving is the existence and development of different modes of organisation of farm production which are linked to the evolution of post-industrial society. This includes most notably the development of part-time and hobby farming in the City's Countryside. The net result of all these various forms is an evolutionary pattern for agriculture in the City's Countryside which is not monolithic but multifaceted. It has been described as the development of a polarised (Bryant et al., 1984) or bi-modal (Smith, 1987) farm structure, although we could argue, based on the growing evidence, for an even more diversified pattern of farm structure (Lawrence, 1988). These different structures should not be taken as the equivalent of the 'dualism' of modern capitalist agriculture, with the development of large-scale, capitalistic units on the one hand and the marginalisation of other units on the other (Buttel, 1982). Indeed, we are suggesting, at least in the City's Countryside, that the different forms of socio-economic organisation of production are all potentially capable of supporting viable alternative life-styles. As Houée (1990, 26) suggests, this multiplicity of forms of production that is surfacing in rural areas shows there is no one model for the future, but rather a 'thousand work sites where the future is being invented'. It is this variety of structures that imparts so much of the resilience to agriculture in the City's Countryside.

SOCIAL AND COMMUNITY PATTERNS. Just as there is growing awareness of the complexity of the structural responses from farming to the various stimuli in the City's Countryside, so also the variety of social structures related to the different farming methods is being recognised (Walker, 1987). Taken together with the different types of people that have taken up residence in the City's Countryside, the resulting communities are anything but simple.

Much concern regarding the transformation of 'rural' communities in the City's Countryside has focused on presumed incompatibilities between the farming population and the newly arrived exurban population (see, for example, Brown, 1977). In turn, these were assumed to reflect fundamental value differences between farming and exurban populations. The incompatibilities are exemplified in the complaints from exurban people regarding 'normal' farming practices such as manure-spreading and the dust and noise of field operations. The presumed importance of such differences has been a major factor in some areas in attempts to keep farming, especially intensive animal husbandry, apart (see Chapter 5 for a detailed discussion on this). The differences in outlook between the two populations have often been assumed to affect attitudes regarding the future of the land. However, research into attitudes reveals a much more complex situation. Smit and Flaherty (1981), for instance, found the differences between farm and non-farm respondents on this issue to be insignificant. Length of residence in the local area, however, turned out to be a much more important factor.

There are also a series of impacts where there is an influx of non-farm population into a community. On the negative side, there is often a transfer of local political power from the farm population to the non-farm population. It has often been assumed that this results in greater demands for more urban services on the part of the non-farm population being realised, with accompanying increases in municipal budgets. We have already mentioned above, however, that differences in attitudes between farm and non-farm populations are not that simple (Smit and Flaherty, 1981), and it is worthwhile noting that the farm population in the City's Countryside shares many of the aspirations for 'better' levels of servicing and material standards of living.

There are also impacts that are more clearly positive. The influx of exurban populations into agricultural communities in the City's Countryside has often been their salvation because of the increased market sizes accompanying exurban development. This has allowed retail businesses to maintain and even expand their operations, creating a vibrant rural service centre. Otherwise, many such centres would have continued to fall by the wayside as commercial service centres because of the continued decline of the agricultural population market and increased consumer mobility (Hodge and Qadeer, 1983).

The salvation of the rural businesses is not always helpful to the farmer, however, and it has been argued that near-urban farmers can suffer as the result of increasing exurban populations if local businesses orient themselves more and more to the exurban market (Beale, 1974; Weeks, 1973). However, generally in such situations, while farm and exurban populations may still function within somewhat different social and economic networks reflecting respectively their longer ties to the local area and the specificity of their agricultural activity (Walker, 1987), what we are observing is an increasing integration of the farm and non-farm communities. This can be expected to be even further advanced in Western Europe.

Not all communities in the vast expanses of the City's Countryside have experienced the stabilising population effects of exurban development. Within the City's Countryside, there are still communities where there is little exurban development, and where there is still agricultural decline. Within the Ile-de-France region in France, for instance, there are still many communities within 40 kilometres (25 miles) of the edge of the Paris conurbation where rural community population has continued to decline, accompanying farm consolidation and release of labour from the agricultural sector (Bryant, 1981). Except where it is associated with a marginal agricultural land resource and farm structure, this is not so much the precipitous decline ('demise' or 'collapse' to use Troughton's terms (Troughton, 1986)) which can be found in some rural areas beyond the City's Countryside, but rather a continued, slow reorganisation of the farm structure and its impact upon the community.

ENVIRONMENTAL RESULTS. Inevitably, these changes in agriculture in the City's Countryside have been linked to the impact on the environment. We have already noted the potential impact of the adoption of modern agricultural technology in industrialised agricultural production upon the long-term productivity of agricultural land. This has caused increasing concern for the sustainability of agricultural production in North America (Brklacich, 1989; Coote et al., 1981; Senate Committee, 1984), the United Kingdom (Munton, 1983b) and other countries (Dahlberg, 1989). Land tenure changes can also exacerbate these impacts, although, as noted earlier, the role of tenure changes is not a simple one (Baker, 1989).

Other effects of industrialised agriculture relate to the increasing scale of intensive animal husbandry and the problems of waste disposal. Together with the off-farm impact of chemical fertilisers, pesticides and herbicides, agriculturally generated pollution has become a major problem in water bodies and rivers. The development of other modes of agricultural production such as alternative agriculture (e.g. Allen and Van Dusen, 1988; Dahlberg, 1986) and less intensive part-time and hobby farming suggests that the situation in the City's Countryside may be less serious than in those areas where industrialised farming is much more pervasive.

Finally, of course, there is the impact on landscape. While it is not difficult to identify the changing landscape features associated with changes in farming (which requires appropriate research to collect the necessary data) (Goodchild and Munton, 1985), it is quite another task to evaluate the significance of such changes from the perspective of land use management and planning because of the subjective nature of the values and images associated with different elements of the agricultural landscape.

4
The farm entrepreneur and the farm

The individual farm level

The importance of understanding the macro environment, including the broad production structure of farming in the City's Countryside, was stressed in Chapter 1. There it was suggested that certain of the structural and behavioural characteristics of individual farms in the City's Countryside are the result of this macro environment and the evolution that this has experienced. Field shape and size, for instance, are partly conditioned by the inherited cadastral pattern, tenure arrangements reflect the existing laws on tenant farming, security of tenure and compensation provisions for tenant investments, and relative price ratios and cost ratios for products and inputs respectively influence enterprise choice and resource allocation on the farm. As the macro system changes, so also do the parameters affecting farming.

Does this mean that the agricultural industry is a passive sector, only responding to external forces of change? This certainly seems to be the perspective taken by many observers of agriculture in the City's Countryside, who have often tended to see farming there as reacting to the forces of urbanisation. However, our earlier discussion of the operation of the forces of change also recognised that the broad forces of change in the macro systems of exchange were also an aggregate expression of individual decisions taken by businesspeople, farmers, households, agencies and groups. This indicates that individual decision-taking units do not have to react to changes going on around them; they can anticipate as well as being more proactive, even to the point of initiating other 'trends'.

Furthermore, while it is unquestionable that farming and farms have undergone significant changes in the second half of the twentieth century, there is a resilience associated with the family farm structure. It has adapted to changing environments, and in some parts of the City's Countryside, farms continue to function in conditions fraught with difficulties for farming. How do we explain this resilience? It is partly explained by an understanding of the broad environment in the City's Countryside in terms of opportunities, a theme that was introduced and developed in Chapter 3. However, we must also understand the role of the individual farmer and farm family in the processes of change around cities.

Our focus in this chapter is therefore on the individual, farm the farmer and, to a lesser extent, the farm family. First, the farm is discussed as a complex operating unit within the macro environment. This sets the stage for a discussion of the role of the individual farmer in decision-taking, and some general frameworks for considering decision-taking are reviewed. An overriding framework is provided by a consideration of the goals and motivations of the farmer and farm family and by the information flows that influence decision-taking. A discussion about decisions on strategies and the key influencing factors provides a partial synthesis of trends and patterns in farming in the City's Countryside, as well as a link to the theme of public intervention and farming in the City's Countryside in the next chapter.

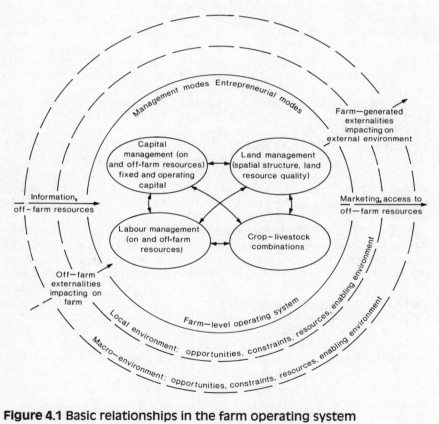

Figure 4.1 Basic relationships in the farm operating system

The farm as a system

Not only does the farm operate within various systems of exchange, but it is also a system in its own right (Morgan and Munton, 1971; Olmstead, 1971). It is common to identify three basic elements to any farming system, i.e. land, labour and capital, and these are usually seen as being integrated into a functioning unit by the management subsystem. It is important, however, to distinguish between management of the farm and entrepreneurial decisions on the farm (Figure 4.1), a point that will be discussed shortly. The specific configuration of the farm reflects a series of evaluations made by the farmer, and farm family, about the available resources both on-farm and off-farm, as well as the opportunities that can be pursued. Changes in the relative values of the different resources considered in the different systems of exchange are influenced by the various forces discussed earlier.

Land is of course a critical dimension and the immediate resource is that which is currently worked on the farm, its spatial configuration (field size and dispersion with respect to the farmstead) and the physical qualities of the individual parcels of land. Many of these are influenced by the macro environment within which the farm is situated, e.g. through the dominant patterns of landholding, laws relating to inheritance and tenure arrangements, land-surveying history and resulting field patterns as well as patterns of non-farm development and highway development.

The potential for change in the land, worked within an individual farm partly depends upon on-farm resources for land improvements and the capacity of the farm to accomodate additional acreage. In addition, change is governed by conditions in both the local and macro environment which influence the supply of additional land for expansion purposes, e.g. interest rates, land prices, supply of land for rental and so forth. Of course, some decisions may be made to reduce the farm's acreage as well as to reduce investment in land improvements. These are discussed later under types of strategies adopted by farmers.

Labour is also a fundamental dimension of any farm. The fact that labour inputs have been reduced substantially over the twentieth century simply reflects the changing values, monetary and non-monetary, attached to labour relative to other agricultural inputs. Labour includes on-farm family labour – the farmer and her or his spouse, and their children – and in the short term at least under some systems, existing hired labour (for some farms, some of their hired labour possess a special relationship to the farm family such that the labour is viewed almost as an extension of the family, and decisions to release such labour are not taken lightly because of a feeling of moral and personal obligation to the people concerned). Off-farm labour includes permanent hired labour, part-time farm labour and seasonal and casual labour. These labour markets are interrelated, and, of course, are influenced by both macro environment considerations such as laws relating to minimum wages as well as local conditions such as local and regional demand for labour in the non-farm sectors. The farmer has to evaluate the demands for labour on the farm, and evaluate the various sources of supply, relative to complementary technologies.

Changes in labour allocation may result therefore from changes in both the macro environment, e.g. changes in inflation rates and labour's expectations of satisfactory income, and in local and regional conditions, e.g. changes in the fortunes of the non-farm sectors. Finally, changes in on-farm conditions can also affect labour allocation decisions, e.g. decisions by farm children to leave the farm and search for alternative employment opportunities.

Capital serves a variety of functions, including operating capital to cover temporary and seasonal cashflow imbalances and fixed capital to secure access to longer-term, fixed assets needed for the farm. For many farms, the business itself is a key source of operating capital, although banks, credit unions, co-operatives and credit arrangements with suppliers are also important. For longer-term capital needs, farmers have increasingly had recourse to off-farm sources of credit, and this has made farms more dependent upon institutional sources of credit, private or public, although family capital remains important for some farms.

Access to credit, and therefore the evaluation and use of capital on the farm – and the resources or inputs to which capital provides access – is influenced by the cost (interest rates) and terms of credit in the macro environment. Governments have played an important role in access to credit. Caldwell (1988), for instance, demonstrates clearly how the general level of overcapitalisation in Quebec agriculture in the 1980s was linked to the provincial government's policy of 'modernising' agriculture. Access to credit is also, of course, linked to the farm's and farmer's ability to carry debt. Thus, the state of financial health of the individual farm and farm family is an important ingredient in the equation too.

The result of the evaluation by the farmer and family of these various on-farm and off-farm resources leads to decisions to continue with, or modify, the enterprise combinations present on the farm. Changes occur frequently in both the macro and local environment of the farm, which may alter the values placed on the different resources, as well as on the rewards to be made from different enterprises. Whether or not these changes lead to changes in the enterprise structure is a function of many factors, including the evaluations and competencies of the individual operator, family influences, and any inertia that exists in the existing farm capital and enterprise structure.

Clearly, all these resource and enterprise allocation decisions are tied together into a functioning system on the farm by making and implementing decisions (Figure 4.2). These processes are influenced by the macro and micro environment within which these decisions are made (Bryant, 1989b). Forces of change in the macro environment (national and international levels) may alter the values placed by individuals on different resources and the relative rewards possible from different enterprise decisions. This macro environment can be defined as an enabling environment for the decision-taker, because to a certain extent its configuration determines the opportunities available to farmers in different regions (through parameters such as price ratios, interest costs, import quotas, supply management, etc.).

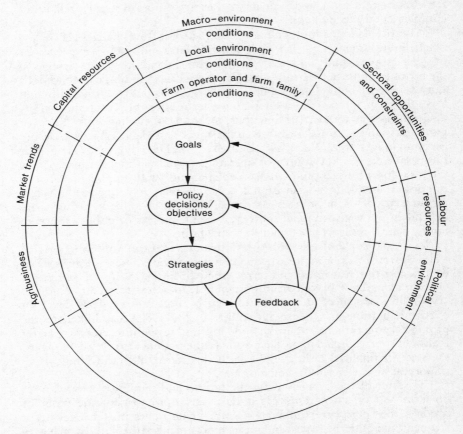

Figure 4.2 Farm decision system and the external environment

Furthermore, this macro enabling environment also influences the extent to which individuals can see an advantage in changing their farm structure. Some enabling environments facilitate change by encouraging innovation and by not penalising efforts to increase revenues, while others seem to be structured to thwart innovation and change. Taxation policies and certain types of national or State/Province 'protective' policies can certainly reduce the incentive, as well as the opportunity, to consider change.

At the more local level there is another layer in the enabling environment, with local planning regulations, for instance, being able to constrain certain types of farm investment decisions through land use controls. This topic is dealt with again in Chapter 5 in more detail. Within the constraints and opportunities set by these enabling environments, there is still considerable room for variation in actual decisions taken by farmers and their families. We therefore now turn to consider decision-taking on the individual farm.

The farmer as manager and entrepreneur

Decisions taken about the farm by the farmer and his or her family can be seen in relation to the achievement of one or more goals, which can be explicitly identified even if they are only implicit in the farmer's and farm family's decisions. These decisions depend on various factors including farm level influences and information flows, both in terms of quantity and quality (Figure 4.2).

Goals and values

One of the reasons for differences in the spatial structure of agriculture is that agricultural patterns result from a multitude of individual decisions (Found, 1971), taken by people who are operating within a wide range of circumstances. Farm decisions are influenced by both internal and external factors. Spatial variation arises when these factors are themselves spatially variable. The goals of the farmer and the farm family represent a key set of internal factors conditioned by influences, both internal and external to the farm. An important question in the City's Countryside is the extent to which there is systematic geographic variation in these goals.

Goals are influenced by the broader values of the society being investigated as well as by the individually held values about expectations by the farmer and farm family. This should warn us that farming in the City's Countryside may be characterised by a different set of decision-making environments because of the potentially different mix of farming structures (see Chapter 1).

Various studies have been conducted into the goals and values of farmers since the 1940s (Clarke and Simpson, 1959; Ilbery, 1985; Kivlin and Fliegel, 1968; Mitchell, 1969; Taylor, 1949). This work took a substantial step forward with Gasson's framework (1973). She identified a list of twenty dominant values 'likely to be associated with the farming occupation' (Gasson, 1973, 527). She then divided the twenty dominant values into four groups representing general orientations to farming:

1. an instrumental orientation to show that farming represents a source of income and security in pleasant working conditions;
2. a social orientation to accommodate those farmers who farm 'for the sake of interpersonal relationships in work';
3. an expressive orientation for those farmers who derive personal satisfaction from or are fulfilled by farming; and
4. an intrinsic orientation to capture those who value farming as an activity in and of itself (Table 4.1).

The results of her study 'suggest that farmers bring a strongly intrinsic orientation to work, instrumental values being rather less significant and social values least so' (Gasson, 1973, 534). Similar conclusions were drawn by Ilbery (1983) in his study of West Midland hop farmers (Table 4.2).

Table 4.1 Dominant values in farming

Instrumental	–	means of obtaining income and security with pleasant working conditions
		making maximum income making satisfactory income safeguarding income with future in mind expansion of business providing pleasant working conditions
Social	–	farming and interpersonal relationships
		gaining recognition and status as a farmer wish to feel part of the farming community carrying on the family tradition working with family members maintaining good worker relationships
Expressive	–	farming as a form of self-expression and personal fulfilment
		pride of ownership self-respect for a worthwhile job using personal special abilities and aptitudes potential for creativity and originality challenging, achieving an objective, personal growth
Intrinsic	–	farming as an activity in its own right
		enjoyment of the work farming as a healthy, outdoor pursuit activity with meaning, value in hard work independence – your own boss Control in varied work situations

Source: Adapted and modified from Gasson (1973, 527)

While there is little reason to think that the general conclusions drawn from the work by Gasson and Ilbery cannot be applied to farmers in general, it is quite likely that a broader range of values would be observed amongst a representative group of farmers in the City's Countryside. This is because many of these areas have a wider range of farmers and socio-economic structures of farming compared with more rural areas. In particular, one would expect a randomly selected group of near-urban farmers to score higher on the intrinsic orientation, reflecting, of course, the greater concentration of hobby and part-time farmers there. Because of such variations, we cannot expect farmers either as farmers or landowners to behave in a uniform manner. Planning intervention that fails to appreciate this will be sorely disappointed (Pyle, 1989).

Table 4.2 Values held by West Midlands hop farmers

Value	Rank[a]
Intrinsic	
Doing the work you like	1
Independence	2
Leading healthy, outdoor life	7
Control in varied work situation	8
Purposeful activity, value in hard work	14
Expressive	
Self-respect for doing worthwhile job	5
Meeting a challenge	6
Using personal special abilities	11
Being creative	12
Pride of ownership	15
Instrumental	
Making satisfactory income	3
Safeguarding income for future	9
Making maximum income	10
Expansion of business	17
Control over hours of work	18
Social	
Earning respect of workers	4
Working with family members, close to home	13
Part of farming community	16
Continuing the family tradition	19
Gaining recognition, status as a farmer	20

Note

(a) The higher the rank (1 highest), the higher the importance placed on a particular value by the farmers. Farmers ranked each value as: essential (4); very important (3); important (2); not really important (1); or irrelevant (0). Scores were then added for each value, and the resulting cumulative scores ranked to give the ranks in the table.

Source: Adapted from Ilbery (1983, 333) and Ilbery (1985, 43).

An aspect of farmers' motivation that has yet to be explored by geographers in any detail concerns the aspirations of other family members, particularly children. However, as Johnston's (1989) work in the area near Metropolitan Toronto revealed, the aspirations of farmers' offspring do play a significant role in the nature of farm-level change. He found, for example, that farmers whose children were interested in taking over the farm were much more likely to undertake changes designed to maintain or improve the economic viability of their operations. Bryant's study (1973b) of the farmers expropriated for the development of the Charles de Gaulle Airport north-east of Paris confirms the importance of these factors. An analysis that attempted to link farmer's decisions with various farm-level influences, including the financial compensation for expropriation, revealed a strong association between the age of the farmer, whether there were children who were taking an active interest in farming (actually working on the farm or enrolled in agricultural courses) and the farmer's investment decisions. Older farmers without children interested in taking over the farm or moving into farming, were content simply to invest their compensation in non-farm investment portfolios. Where there were children interested in taking over the farm or working in agriculture, the farmers regardless of age, became involved in all kinds of farm investment strategies, including intensification on the part of the farm remaining after expropriation and the purchase or lease of farms in other regions for the eventual relocation of the farm (in anticipation of further expropriation) or simply to provide opportunities for immediate family members to enter farming.

These studies suggest the importance of understanding the whole family unit in order to unravel farming decisions, and logically lead to a consideration of the family and farmer life-cycle to summarise these micro-level influences (Bryant, 1986a; Moran, 1988; Munton et al., 1988). Moran, for instance, suggests a link between the family cycle, the equity that farm families have in their farm property and involvement in off-farm work. As the farm family ages, their equity progresses from one of high debt to one of low debt, if any at all, after twenty or thirty years. This he suggests is sufficient to generate a range of responses from farmers and their families to the same pressures, on top of other variations related to inherently psychological attributes or variations in the macro-economic environment. For two samples of dairy farms in New Zealand's North Island, Moran demonstrates a strong positive relationship between the farmer's age and equity in the farm property, which implies a direct link as well with income because of the link between equity and interest payments. The analysis suggested that farmers with lower debt, and higher equity in their property, were more likely to invest in less labour-intensive systems of production.

Another example of the relevance of considering farm change from the perspective of the whole family is hinted at in Bryant et al.'s (1991) and Marois et al.'s (1991) analysis of the behavioural considerations in farm change in the Toronto and Montreal urban fringes. In these study areas, farmers were significantly older in the most urbanised areas compared with the remainder of the study areas. This would be easy to interpret as reflecting the lack of successors to take over the farm, an indicator of degeneration (and, certainly, this is part of the explanation). However, in the Montreal area, there was a much greater proportion of farmers who reported a previous occupation before taking over the farm. Since these were not part-time farmers, one explanation is that the smaller farms in the Montreal area are less able to support two families. Hence, younger would-be farmers spend time off the farm early on and only take over the family farm when the older generation is close to retiring completely from agriculture.

In the final analysis, of course, we have to put these micro-level influences and variations into the broader context, since this macro environment also sets broad parameters within which farmers and their families take decisions.

The importance of understanding the individual farm-level processes is, however, twofold:

1. it shows that there is indeed a fundamental heterogeneity of response within the farm system, even within a farm community characterised by a homogeneous farm structure; and
2. having once recognised this heterogeneity, there are important implications for the way in which public intervention addresses certain issues (see Chapter 5).

Changing goals and the urban world

Much has been written about the differences between farmers and non-farmers in terms of goals, values and aspirations. In the City's Countryside, where these two populations have the opportunity to come into close contact with one another, the situation appears to be ripe for conflict.

The differences in socio-economic status of farmers and their non-farm neighbours in areas near cities are well documented. Most studies of exurban development have found the new, incoming population to be younger, more affluent and better educated than the indigenous population (see, for example, Campbell and Johnson, 1976; DeJong and Humphrey, 1976; Graber, 1974; McRae, 1981; Mitchell, 1975; Ploch, 1978).

The conventional view is that the non-farm population also hold values and attitudes which cause them to come into conflict with their farm neighbours (Smit and Flaherty, 1981). However, while there are cases where value differences between these two groups have been documented (McRae, 1977; Walker, 1977), there is also evidence to suggest that such differences may have been overstated. For example, Smit and Flaherty (1980) found no difference between farmers and non-farmers in terms of the opinions on local severance policy, and Joseph and Smit (1981), in an examination of preferences for municipal services, could find no evidence that the preferred service allocations of farmers and non-farmers were significantly different. In his study of farming in the area just north of Metropolitan Toronto, Punter (1976) discovered that non-farmers were often supportive of their farm neighbours, a finding which can be explained by the fact that many exurbanites are drawn to the City's Countryside because they find the landscape an attractive one to live in.

Indeed it is easy to think of a whole host of mutually beneficial arrangements being struck between farmers and non-farmers in the City's Countryside (Bryant and Russwurm, 1979). It is acknowledged that conflicts do occur and that the cause of the conflicts may well be different values, but then few of us can claim to be on good terms with all of our neighbours. Why should farmers and non-farmers in the City's Countryside be any different? Even if there were found to be significant differences in values between farm and non-farm groups in the City's Countryside, it is difficult to see how this could be used as an argument to contain the development of one population group compared with the other.

Land ownership and security as goals

The difficulties associated with trying to draw clear distinctions between farmers and their families, on the one hand, and the non-farm population on the other is brought into sharp focus when we consider landownership and its well-entrenched role in our society. The ownership of farmland can be seen as desirable by many farmers because it provides security for the future. Because of its importance and because of the increased difficulties of achieving this in parts of the City's Countryside (see Chapter 3), it could be argued that farmland ownership by farmers is a significant indicator of agricultural 'health' in the City's Countryside. But is it as simple as that?

Farmland ownership provides more than just access to the productive farmland for the farmer: if that were the only reason, then renting farmland would make more sense economically, providing adequate security of tenure was assured and compensation available for tenant investments made in the land. However, ownership of farmland involves a number of other rights as well, including that of being able to sell the land and realise any appreciation in capital values. For farmers, who are in an industry in which the cost-price squeeze has frequently led to very low levels of return to management and capital invested, the option of one day selling off part or all of their land for purposes other than agricultural production is a very attractive one, especially of course in the inner parts of the City's Countryside. This means that farmers who are owner-occupiers wear two 'hats', that of farmer and that of landowner. The dilemma, of course, is that the two do not always converge, and farmers have often been among the most vocal of those opposed to programmes of farmland conservation.

Most farmers do not think of themselves as land speculators. Some argue that farmers are forced to be land speculators because the equity they hold in their business is often their only security for the future and land is generally their single most important asset. As we will see in the next chapter, research has shown that most farmers believe in maintaining the integrity of agricultural land. However, at the same time, most want to be free to sell their land to the highest bidder. This is perfectly understandable, since for most farmers their land is also their retirement fund.

Information flows and farming in the City's Countryside

Farmers in near-urban zones certainly face some unique problems and some farmers find these problems insurmountable. However, this tells only part of the story. There is simply too much agriculture in areas near cities for us to accept the pessimistic view that agriculture in the City's Countryside is on an inevitable downhill path.

We have already argued in Chapter 3 that near-urban farmers derive some benefits because of their location. In that chapter, the emphasis was placed upon market opportunities afforded by the nearby urban markets. Another benefit relates to the volume and quality of information to which near-urban farmers have access, an idea which figured prominently in Schultz's (1951) urban-industrial hypothesis of agricultural development. Despite the importance of information, however, especially for decision-making, the adoption of innovations (Jones, 1963, 1967), and the resultant impact on agricultural landscapes (Found, 1971), it has not been a significant topic of inquiry by agricultural geographers (Ilbery, 1985).

It is important to recognise the difference between sources of information and the use of information (Pred, 1967). Pred's 'behavioural matrix' was devised to try to account for decision-making behaviour on the basis of information quantity and quality on the one hand, and the farmer's ability to use the available information on the other. Where there is a high level of competency on the part of the farmer and a high quality and volume of information available to the farmer, the argument is that a sound 'economic' decision is likely to be taken. To a certain extent, this framework is an oversimplification of behaviour (Harvey, 1969), because the two dimensions of quantity/quality of information and ability of the decision-taker are in fact highly interrelated 'as the information sought by the decision maker and hence "available" depends on his [sic] perception of the environment, which in turn is related to his ability' (Morgan and Munton, 1971, 35). None the less, it suggests that if information is richer in the City's Countryside, then 'better' economic decisions are more likely, all other things being equal.

Morgan and Munton (1971) classify the various sources of information to which farmers have access into two groups: (1) sources external to the agricultural society, e.g. agricultural advisory services, research centres and mass media sources; and (2) sources within the agricultural society based on interfarmer personal contact.

They also make the point that most farmers look upon information from sources outside the agricultural community with a degree of suspicion. Farmer's mistrust of government is almost legendary. What student of agricultural geography conducting farm interviews has not been regarded, at least initially, with suspicion of working for the 'government'? Still, on the other hand, Jones (1976) contends that because each farm is independent, a business unto itself (which means that all farmers are basically competing with one another), farmers often conceal information from one another.

In so far as the question has been addressed, farmers in near-urban zones are at an advantage over farmers elsewhere in terms of access of information. Not only do farmers in the City's Countryside have the opportunity to read the same farm magazines and listen to the same farm broadcasts as other farmers, but those near cities also have better access to various other information sources. Many of the major agricultural research centres are located in or near large towns or cities, as are most universities which are one of society's major generators of information. The urban-industrial complex is still the location where most innovations take place, and whether or not agricultural application is the intent of the inventors, near-urban farmers enjoy a comparative advantage with regards to access to this information.

In addition to the provision of information influencing agricultural change, researchers have also focused on the willingness and ability of farmers to use information (see, for example, Jones, 1963, 1967). Generally speaking, farmers who are young, well educated and who run large, intensive, specialised operations are more likely to seek out and use information than farmers who are older, less well educated and who operate on a relatively small scale (Ilbery, 1985). This suggests a number of fruitful hypotheses that could be pursued in investigations of farm change in the City's Countryside.

We have already noted the potential significance of the mix of different farming systems in the City's Countryside, raising the likelihood that different types of information are important to different business structures. In the context of innovation, an interesting hypothesis links the hobby-farmer as an agent of change. If, as Morgan and Munton (1971) suggested, (some) hobby farmers are generally more concerned with technical efficiency than with profit maximisation, it is possible that this group could play an important role in the introduction of new ideas into the farming community. In field studies by Johnston (1989) near Toronto, several of the farmers interviewed commented that they adopted production ideas first tried out by their hobby-farmer neighbours. Such evidence is, it must be stressed, anecdotal and impressionistic and, of course, requires extensive empirical verification.

Types of farm decisions

Farmers operate therefore in a complex and dynamic decision environment according to various goal sets and information they receive that allow them to evaluate on-farm and off-farm resources, constraints and opportunities. There are many types of decisions in which farmers and their families become involved, and there are several important distinctions that will help place our discussion of farm decisions in the City's Countryside into proper perspective.

One important distinction is between policy decisions and strategic decisions. The former are decisions which affect the essence of an operation and which establish its broad parameters, whereas strategic decisions outline the means to implement the policy. In addition, it is also important to separate the day-to-day operational decisions, e.g. whether to harvest today, whether to allocate labour at a particular time to one enterprise or another on the farm and so forth (Morgan and Munton, 1971). Obviously, these types of decisions are highly interrelated. For instance, problems encountered in the day-to-day allocation of labour to different tasks on the farm may signal an error in making a strategic decision about enterprise combinations or in the allocation of capital to labour-saving technology.

As an example of the difference between policy and strategic decisions, we can cite the situation in New Zealand, starting in about 1985, when many farmers were affected by what has come to be called the Rural Downturn. Within a period of several months, a central government price support scheme was withdrawn, prices for a wide range of farm products fell dramatically and interest rates climbed from around 13 or 14 per cent to over 20 per cent. As a result, many New Zealand farmers found themselves with very serious cash flow problems, particularly in meeting their interest payments. A number of farmers interviewed in the lower North Island in connection with research into the changing structure of New Zealand farms, reported taking the (policy) decision that their operation needed to be restructured in order that incoming cash would flow more evenly throughout the year (Johnston, 1990). A number of strategic decisions followed. For example, some sheep farmers added beef cattle to their operations, which can be marketed at a different time of the year than sheep, lambs and wool. Similarly, Morris (1989) found that apple-growers in the Hawkes Bay area (a major fruit-growing region in the North Island) had begun to plant a much wider range of apple varieties than was the case previously. Moran et al. (1989) report increases in multiple job-holding within a sample of hill country sheep farms. All are examples of strategic decisions taken in the context of a broader, more general policy decision to increase the security of the cash flow into the farm and farm family.

Another way of looking at farm decisions is to differentiate a management mode of decision-taking and normal management decisions on the one hand, from entrepreneurial decision-taking modes and decisions on the other (Bryant, 1989b). A management mode of decision-taking involves the farmer working essentially with familiar production techniques, enterprise structure and marketing channels. By contrast, an entrepreneurial mode of decision-taking involves more fundamental changes to the farm as well as the incorporation of entirely new techniques, enterprises and marketing approaches. It is one that is associated with a higher degree of risk than management-type decisions. The entrepreneurial farmer is a decision-taker who can identify an opportunity, pull together the necessary factors of production and turn the idea into practice (Grasley, 1987). Characteristically, they are people who go beyond the immediate resources over which they have direct control. Examples of entrepreneurial decisions might include adopting an enterprise entirely new to the farm or, indeed, the local area and developing a new form of marketing for the farm's products. Specific examples are identified later in this chapter.

How is the decision-making process structured?

Various social theorists have developed models to describe or explain the way in which decisions are made (see, for example, Faludi, 1973a, 1973b). Although these models were not specifically designed to describe or guide decision-making at the level of the individual farm business, their translation into that context can be undertaken in a relatively straightforward fashion. Three decision-making models, which each represent very different ways of approaching decision-making are: (1) the rationalist approach; (2) the incrementalist approach; and (3) the mixed-scanning approach.

The rational model, which, according to Faludi (1987) was introduced into the planning literature by Meyerson and Banfield (1955), is the most clearly normative of the three reviewed here. Briefly, faced with a choice it is suggested that the decision-maker will conduct a comprehensive and exhaustive analysis of the problem at hand, identify and carefully assess the widest possible range of ways of addressing the problem, and then identify the 'best' solution (in relation to a defined set of goals and objectives). The rationalist model has obvious parallels with the notion of 'economic man'. Critics of the rational model abound (see, for example, Faludi, 1987), typically arguing that it fails both as a normative model and a descriptive model because the demands of the approach are generally beyond the capacities of decisions-makers and the structures within which they operate (Braybrook and Lindblom, 1963). Information to make decisions is often incomplete, largely because the costs of exhaustive information collection are prohibitive. Even if a comprehensive information collection effort is made, synthesising and analysing large volumes of information are beyond the majority of individuals and organisations, particularly if the information is variable in form (e.g. some quantitative, some qualitative) and detail (e.g. different spatial scales). On the other hand, the rational model does provide a useful conceptual yardstick against which to assess decision-making.

In reaction to his dissatisfaction with the rationalist model, Lindblom (1959) offerred the disjointed-incrementalist approach to decision-making. Less demanding than the rational model, Lindblom (1965) suggested that decision-makers focus only on policies that are incrementally different from existing policies, and that typically only a small number of policy options and their consequences are considered. The problem being addressed is continually being reassessed, and the incrementalist approach allows for adjustments between goals and objectives and strategies. Viewed in this way, incrementalist approaches to decision-making seem almost remedial, more in line with the management mode of decision-making and management decisions noted in the previous section than with the entrepreneurial approach to decision-making. It is also easy to see how such an approach might fit in well with notions of 'satisficing' behaviour, as opposed to maximising behaviour, and how it readily accommodates the notion of inertia in farm structures.

As an alternative to both Lindblom's model and the rationalist model, Etzioni (1973) offered the mixed-scanning approach. Arguing that this model is both an accurate description of the way in which many decisions are made as well as an ideal, the mixed-scanning approach follows two general steps. First, once a problem has been defined, a general assessment of the entire situation is undertaken. In this step little detailed information is collected but a comprehensive listing of options is undertaken. The second step involves close, detailed examination of a small number of promising options. For instance, let us say that a farmer feels the need to increase farm revenues. Using the mixed-scanning model, the first thing the farmer does is prepare a list of all the possible ways to increase farm revenues which would be compatible with his or her goals. The next step is to narrow the possible options to a workable number (the number considered will depend heavily on the time and resources allocated to the process), and detailed information collected and analysed. It is an approach that appears to accommodate both managerial and entrepreneurial modes of decision-making, since these simply affect the range of policy and strategic options that are considered.

At this point, it is worth emphasizing that while we have referred frequently to the farmer as decision-taker, manager and entrepreneur, farm decisions in the context of the various types of family farm are not necessarily taken solely by the individual farmer. We have already stressed the importance of the farm family's influence on decision-making. In addition, it is important to acknowledge the role of the spouse and children in contributing directly to the taking of decisions on the farm. This role varies substantially between farms (Buttel and Gillespie, 1984).

On farms where more sophisticated management structures have been adopted, particularly in more capitalistic operations, two characteristics are worth noting. First, we find that in such operations attempts are made to identify clearly the different areas of decision-making, and assign the primary responsibility for them to particular personnel depending upon their skills and interest (Moore and Dean, 1972). This division of responsibilities certainly occurs in more traditional systems. The difference in a more capitalistic and/or industrialised operation is that assigning tasks to specific people tends to be an imposed structure of management rather than one which is simply allowed to evolve. The second noteworthy development is an increased reliance on external experts who possess very specialised knowledge (Errington, 1986). Livestock farmers have for years relied upon their veterinary surgeon, for example, but increasingly many other agricultural specialists have come on the scene. This development is not surprising because farming has become more and more complex over the years and it is beyond all but a few farmers to possess knowledge and skills required to run a successful operation today without any assistance. This increased reliance on external knowledge sources and skill pools is really just part of a more general set of changes which have swept over agriculture in the developed world, as changes in farm production have brought in their wake a whole new set of systems of exchange.

Farm decision-making in the City's Countryside: a synthesis

In order to synthesise the various dimensions of farm decision-making already introduced, a conceptual framework is developed below. First, the framework introduced initially in Chapter 1 (see Figure 1.3) is expanded to focus upon decisions taken at the individual farm level. Second, a framework of the various strategies that farmers can take in the context of the pressures and stimuli facing them in the City's Countryside is presented. Third, a geographic synthesis of farm and agricultural change in the City's Countryside is briefly given.

A conceptual framework of farm adjustment in the City's Countryside

The conceptual framework illustrated in Figure 4.3 was developed by Johnston (1989), and in turn based upon work by Olmstead (1970), Bryant (1976, 1981, 1984a, 1986a) and McCuaig and Manning (1982). Three major concerns are addressed in this framework:

1. the need to recognise as broad a range as possible of the major influences affecting farm change;
2. the need to be able to accommodate a wide range of types of farm change; and
3. the need to identify the major relationships and linkages between the basic elements of the framework.

It is assumed that farm-based changes represent responses to some stimulus or stimuli; farm change is therefore seen as a response to an alteration in the farm's operating environment by which the values attached to the inputs and products that are part of the farm's actual and potential system of exchange are modified. Such changes in the farm's operating environment can be either: (1) actual changes that have occurred or are occurring; or (2) changes that the farmer anticipates; or (3) changes in values that the farmer attaches to elements in his or her operating environment. The framework thus covers reactive and adaptive responses, the latter including entrepreneurial responses.

Farm change is influenced by both external and internal factors (Bryant, 1974, 1976; Munton et al., 1988; Olmstead, 1970). Focusing first on the external factors, two distinct types are recognised, as noted in Chapter 1. In the City's Countryside one type is represented by metropolitan-based factors or influences which relate in some way to urbanisation, urban expansion or exurban development. Specific influences which would be thus categorised include the demand for land and labour for non-agricultural purposes and the market for agricultural produce represented by urban areas (see Chapter 3). Non-metropolitan factors, on the other hand, refer to influences not directly related to urbanisation. There are influences which Munton (1974) argued researchers tended to underestimate and even sometimes ignore: they include such things as changes in interest rates, commodity prices, national-level 'farm' policy and also societal values and aspirations. We also

THE FARM ENTREPRENEUR AND THE FARM

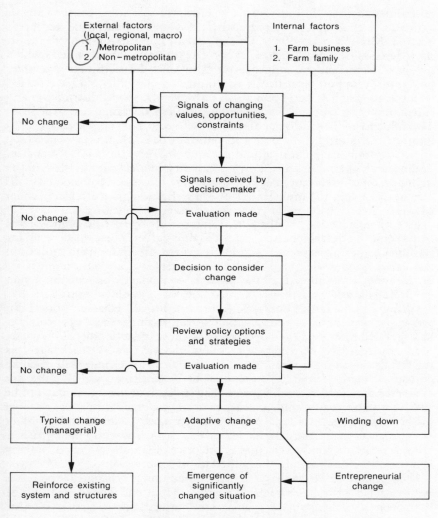

Source: Adapted from Johnston (1989; 58); Bryant (1989b)

Figure 4.3 Conceptual framework for farm adjustment in near-urban areas

We also have to recognise (Chapter 1, Figure 1.3; Bryant, 1984a) that agricultural and farm changes operate in a particular regional environment, with specific economic, political and biophysical characteristics, which can exert a considerable influence.

It is also important to note that these forces operate on different levels. Many of the non-metropolitan factors operate through systems of exchange on the macro scale. In this case we may expect that some components of agricultural change will be shared by the City's Countryside and areas beyond (Munton et al., 1988). One important non-metropolitan factor on the macro scale involves the role of the state through various policies which can have a tremendous influence on farm-level decisions. For example, government support of particular agricultural commodities is often reflected in enterprise choice. Fiscal policy can influence the spatial distribution and extent of specific types of farming because it affects operating cost structures. The influence of the state is not limited to these examples, but they do serve to illustrate the point that the state's actions must be taken into account in any attempt to understand agricultural patterns. Indeed, too many researchers dealing with agriculture in the City's Countryside relegate the state to the unimportant or irrelevant category. When this dimension is considered at all, it is in discussing the implications of research findings, particularly when advocating centralised public intervention in farmland conservation.

Metropolitan factors, on the other hand, usually operate through systems of exchange on the regional or local scale, although of course the demand for labour from non-farm sectors is widespread in its effects. Many metropolitan factors are quite localised and display considerable variation from one near-urban zone to another, e.g. the patterns of demand for exurban living environments and growth pressures around a metropolitan centre.

Changes in the farm's operating environment from whatever source can warn the farmer that a re-evaluation of the resources and opportunities incorporated into the farm's operating system may be in order. Of course, owing to differences between actors, such as the way in which each perceives the farm's operating environment, not all signals for change will reach all decision-makers. Similarly, reception of a particular signal does not guarantee that a change will happen. The strength of the signal may not be sufficiently strong to stimulate a change or a farmer may simply be too set in his/her ways to undertake certain changes. Alternatively, other circumstances may reduce the likelihood of a change, as with a hobby farmer who does not respond to changes in the price of labour-saving technology because of the time already committed to off-farm work.

While a signal for change is a necessary condition for change to occur, it is not by itself a sufficient condition for a change to occur. The farmer must be both open to the possibility of change and have the ability to carry it out. One of the negative consequences, for example, of the industrialisation of agriculture is that commitment to particular production systems is often so great that the farmer's ability to respond to changing structural conditions can be inhibited. Once farmers get on the technological treadmill, it is extremely difficult for them to get off. Given a decision to make a change, then it is time to consider what options are available. The distinction between these two stages will obviously be clearer for some farmers than for others; not everyone approaches problems the same way (see the brief discussion above on different models of decision-making). Those two steps are also often carried out in an iterative fashion. If, for example, the initial consideration of possible courses of actions fails to yield an acceptable one, then the decision on whether or not to make a change at all may be reassessed.

Policy decisions and strategies (which are dealt with in more detail below) can be categorised as being either typical change, adaptive change or winding-down choices (Bryant, 1989b). Only the first two are commented upon here. Typical changes comprise adjustments which are not situationally or site specific and can be observed on a large number of farms. Adjustments such as increasing farm size or the substitution of capital for labour are examples of changes occurring throughout agriculture and hence are typical. Adaptive adjustments, on the other hand, are more specific than typical changes, and are implemented specifically to exploit, or respond to local or unique circumstances. Adaptive adjustments should therefore be observed on fewer farms in any given region than typical adjustments. Typical changes reinforce existing geographic patterns, whereas adaptive changes alter existing patterns in some fundamental way. Clearly, however, adaptive change over time can merge into 'typical' change when those who engage in adaptive change are actually the early adopters in its broader diffusion.

Most typical changes, given that they involve doing things that have been done before, are relatively low risk adjustments and therefore fall nicely into the category of management decisions identified earlier in this chapter. This means that the knowledge and ideas which underlie such changes are generally well established. Adaptive changes, on the other hand, are subject to much more risk; the ideas being applied in adaptive change are typically new and untried. Farmers are often unaware in any detailed sense of how the adjustments will turn out. Those implementing adaptive changes are generally entering unchartered waters. While adaptive strategies cover a fairly wide range, including moving into part-time farming in some areas, many of the adaptive changes are in fact entrepreneurial. It is to this area that we now turn our attention briefly.

Entrepreneurial spirit and the City's Countryside

From the earlier discussion of the difference between managerial modes and entrepreneurial modes of operating, entrepreneurial activity must be understood in the context of: (1) the macro and micro-enabling environment; (2) the perceived need to undertake decisions that are out of the ordinary (stress); and (3) the individual farmer's and farm family's abilities and personal inclinations.

Farmers are frequently faced with taking decisions in an environment of risk and uncertainty, simply because no one, farmers included, knows the future course of events for certain. Prices, costs, weather, yields, political decisions, family circumstances and many other variables are all subject to change and therefore influence the future rewards of taking decisions now. These are all forms of risky situations that are created by factors largely beyond the control of the farm. In response to these situations, some farmers purposely diversify production so as to spread risk rather than concentrate it, others take on enterprises that are less likely to experience erratic returns, and still others take out various forms of 'insurance' (Found, 1971).

These risky situations exist in the City's Countryside as elsewhere, and farmers' responses to them will vary according to the range of factors normally affecting farm decision-making which were discussed earlier. However, with adaptive, entrepreneurial change, some farmers enter into risky situations by choice, and there are at least three reasons why we might expect to see more entrepreneurial spirit in the City's Countryside.

First, because farmers in the City's Countryside are subject to the same sets of forces as farmers elsewhere plus the forces that are unique to near-urban zones, they have a larger set of entrepreneurial opportunities open to them. Many of these can be linked to the evolving needs of post-industrial society and developments in personal mobility. These combine to yield a greater intensity of opportunity in the City's Countryside than elsewhere (Laureau, 1983; Lockeretz, 1987; Mainié and de Maillard, 1983). Thus, the opportunity to sell products directly to the customer is bigger in near-urban zones simply because this activity depends upon population thresholds that are not achieved in more remote areas. Other benefits are derived from direct marketing which, in turn, can foster entrepreneurial activity. Farmers who are part of the traditional agro-commodity chains never have personal contact with the final consumer. Consumer preferences, such as for leaner beef, are therefore filtered through several different agents and along the way can be distorted. These agents can, for example, alter the message to reflect their own agendas. Even if the message from consumers that does reach the farmer has not been modified, there will still be a delay between its expression and its reception. However, farmers who have direct contact with the consumers are in the enviable position of being able to gauge consumer preferences and make adjustments directly (Moore, 1990). The activities involved in these entrepreneurial endeavours are, of course, usually outside the domain of state-regulated enterprises.

Closely linked with the notion of better market opportunities is an enhanced potential for innovation in agricultural technologies in the City's Countryside. In the early stages of urbanisation and industrialisation, this led to the hypothesis that links the rate of change in agriculture to the degree of integration of agriculture with the urban-industrial complex (Schultz, 1953). The urban-industrial complex in this early period was a major source of technical innovations in agriculture, particularly in terms of labour-saving technologies which had strong links with industries such as the machine-tool industry (Pautard, 1965). Subsequent developments in information communication and personal mobility has removed the more obvious impediments to the diffusion of information, so that it may be acquired by agricultural areas anywhere in the developed world. None the less, the search for information and the willingness to adopt innovations are important ingredients of the equation. Since farmers in the City's Countryside have already been exposed to more contact with the urban-industrial complex and with the different life-styles of the urban and exurban populations, and have made adjustments, then they might also be expected to be more willing to contemplate change.

The second reason for more entrepreneurial activity in near-urban zones is because many of the commercial farmers operating there have had to be adaptable. The greater stresses in parts of the City's Countryside have created an incentive for farmers to be innovative (and the opportunities present have allowed many of them to pursue this non-traditional form of adaptation). As Lockeretz (1987: xxii) expressed it: 'Out of necessity metro farmers have come up with new ways of making do with what they have.' 'Difficult circumstances sometimes are just what is needed to stimulate new thinking when old ways of doing things no longer work.' Thus, we should expect to see a stronger entrepreneurial spirit amongst farmers near cities because many of them have had to be entrepreneurial to remain in production.

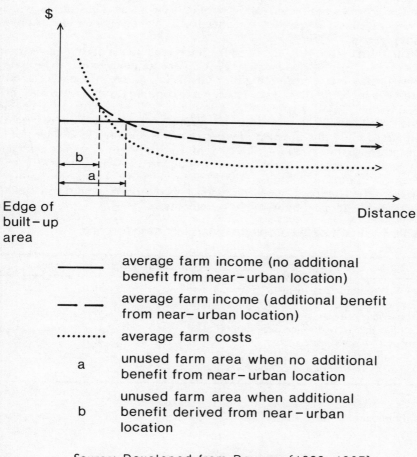

Figure 4.4 Effects of additional income benefits from near-urban locations on the setting aside of farmland

This phenomenon has been suggested as a partial explanation of the fact that there appears to be less setting aside of farmland in areas of urban expansion than one might otherwise anticipate (Dawson, 1982, 1987) (Figure 4.4). Dawson (1982) argued that owing to such things as increased property taxes and cost associated with trespassing and vandalism, farmers' production costs in near-urban areas are typically higher than their counterparts elsewhere. Dawson suggested that farmland would be set aside at the point where increased production costs exceed revenues (Figure 4.4). However, in a subsequent paper Dawson (1987) observed that the setting aside of farmland is not as extensive as some theorists have suggested (see, for example, Sinclair, 1967). The explanation he offered is that faced with climbing production costs, some farmers have made adjustments which boost revenues (Figure 4.4), effectively by engaging in adaptive, entrepreneurial activity.

The third reason that we see more entrepreneurial activity in areas near cities is because of the wider range of socio-economic structures in farming. In particular, it is worth noting the potentially positive influence of hobby farmers. Hobby farmers, who are known to concentrate in near-urban zones (Blair, 1980;; Layton, 1976, 1978; Troughton, 1976a), are often the people who introduce new ideas into the local farming community. Hobby farmers can afford to do this since maximising profits is seldom one of their prime concerns. Gasson (1968) found, for example, that hobby farmers often rank economic considerations below such concerns as farm appearance and technical 'optima' (e.g. high yields). New production systems, for example, are often tried out by the non-commercial operators first, and the successful ones quickly adopted by watchful neighbours.

Table 4.3 Urban-based problems of farmers near Toronto

Problem	Number of farmers citing problem[a]
Local community has changed	40
Too much traffic	39
Trespassing and vandalism	26
Complaints from non-farm neighbours	20
Prospective landfill site nearby	12
Taxes are too high	8
Too many restrictions on farming activities	6

Note

(a) Number of farms in survey was 189.

Source: Johnston (1989)

Strategies for farmers in the City's Countryside

There can be little doubt that farmers in the City's Countryside are faced with some unique urban-based pressures (Table 4.3), and certainly these problems are sometimes so severe that some farmers find it impossible to remain in business and others are only able to get by. However, the opportunities for some farmers to engage in entrepreneurial activity or otherwise to adapt positively to these pressures means that we also have to recognise many other types of policies and strategies.

The principal types of policy options and examples of the types of strategies involved are shown on Figure 4.5. Many of these have already been introduced and discussed elsewhere in this book. Here, only the broad framework is outlined. The policy options deal essentially with decisions to expand or contract the farm business size, to shift labour partly or wholly out of agricultural production and to modify the organisational structure of the farm, which essentially involves altering the structure of capital ownership. The first two policy options basically include modifications to the volume and ratio relationships between the different factors of production used on the farm. Choice of policy option, and, of course, specific strategies is influenced by all the various considerations that we have already discussed. Hence, we shall comment only on a selection of the options here, rather than discussing the formative factors and circumstances. Strategies can be any combination of 'typical' changes and adaptive changes, the latter including entrepreneurial changes. Furthermore, the strategies in particular are not mutually exclusive, and it is possible of course to observe combinations of different strategies, e.g. farm business contraction combined with the decision to move into off-farm employment and part-time farming.

The decision to expand the farm can be implemented through a number of strategies. The most obvious one is expansion of the physical acreage of the farm (Bryant, 1976; Munton et al., 1988), either through rental of more land or purchase. Rental of additional farmland has been particularly important in farm amalgamation in North America (Fielding, 1979; Sublett, 1975) because of the relative ease with which rental agreements can be entered into. However, there has always been a strong tendency for farmers to want to own their land; indeed, in the United Kingdom, the proportion of owner-occupied farmland has increased in the 1970s and 1980s while that of rented ('let') land declined markedly (Marsden et al., 1989; Munton et al., 1988).

This whole process of farm size expansion, whether through rental or by purchase, has been driven by the economic and technological imperatives to achieve economies of scale in modern farming. In the City's Countryside, farmland rental has often allowed this to occur quite easily and cheaply, notwithstanding concern that has often been expressed about the quality of management applied to rental land. The potential for this to be exacerbated appears higher where the leases are insecure, as often happens in the inner urban fringe. (Munton et al. (1988) note, for instance, the substantial increase in 'insecure tenancies' in their sample of farms in London's Metropolitan

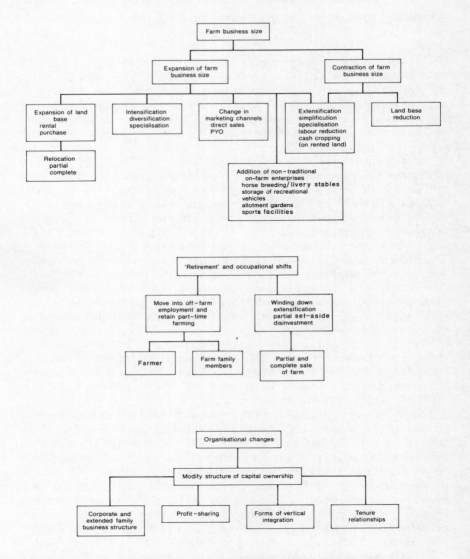

Figure 4.5 Policy and strategy options at the farm level in the City's Countryside: key examples

Green Belt from 2.8 per cent to 6.1 per cent of the land area between 1970 and 1985.) Munton et al. (1988) also report that eleven of their sample of eighty-five farmers in London's Metropolitan Green Belt also had active land policies, by which they bought and sold land in order to 'restructure their land holdings'. This appears to have been undertaken not so much to increase the size of the landholding as to manage the farm debt load more effectively.

Another way of expanding the acreage of the farm is through the purchase (or rental) of farmland quite separate from the original farm, or even another whole farm. In some cases this is undertaken to prepare for a partial or complete relocation of the farm business or simply to permit the sons and daughters of the farm family to enter farming. In most farm surveys in the urban fringe, this form of adaptation does not stand out in terms of magnitude. However, where the pressures on a whole farm community are extreme, this might be expected to be more common. Thus, Bryant (1973) in an analysis of farmer decisions consequent upon expropriations for the Charles de Gaulle Airport north-east of Paris, France, found that acquisition of a farm in a new location for either their own or their children's eventual relocation was a major form of adaptation to the actual and perceived future pressures on farming in this urban fringe area. In many cases, the actual amount of land that came under the control of the farmers engaging in this form of adaptation increased substantially after expropriation! This reflects the higher land prices (and, in this case, the high compensation rates for expropriation) in an urban fringe. Lucas and van Oort (1991) also report in their sample of farmers who had sold land near Utrecht that 30 per cent had decided to resettle elsewhere.

The farm can also be expanded through intensification, thereby increasing the production of the business; this frequently involves more capital investment. Intensification can occur in association with changes to the degree of agricultural enterprise diversification within the farm and even along with contraction of the acreage of the farm. On the one hand, some studies have seen an increase in agricultural diversification within particular farms, involving both traditional (e.g. Deslauriers et al., 1990) and non-traditional enterprises (e.g. Gilg, 1990; Ilbery, 1990). In other cases, however, intensification has been associated with increased specialisation within the individual farm (e.g. Bryant et al., 1989). Different diversification strategies through the net addition of agricultural enterprises, non-agricultural enterprises or alternative marketing channels all seek to increase the number of different markets in which agricultural resources bear fruit. Diversification, assuming the ultimate markets are in fact different, reduces dependency, stabilises income (here, diversification plays the role of one of Marsden et al.'s (1986a) survival strategies) and can even increase income (and therefore become part of an accumulation strategy) (Deslauriers et al., 1990). Diversification is therefore not necessarily a weakness, as has so often been assumed under the industrial model of agricultural production.

Intensification of agricultural production is often associated with another strategy, that of modifying or developing entirely new forms of marketing channels. As discussed in Chapter 3, the City's Countryside is a favoured location for various forms of direct marketing of farm products. There is substantial variety in the products and the forms that direct marketing takes, including farm shops and pick-your-own operations (Bowler, 1981a, 1981b; Laureau, 1983; Rickard, 1991). In Ilbery's (1988b) study in the Birmingham, United Kingdom, urban fringe, of the 120 farms with 'alternative' enterprises in the sample, direct marketing was the most frequent form of diversification (82 of the farmers).

In Johnston's study (1989), 31 of the 189 farmers surveyed near Toronto had become involved in direct sales to the public (Table 4.4). A wide range of items were sold direct to the public including livestock (sheep and lambs, beef, rabbits, poultry and eggs), various nursery products, and fruits and vegetables. In many cases, these are products whose production is not regulated by such agencies as marketing boards, and thus fit nicely into an entrepreneurial mode of operation. All kinds of selling strategies were observed, ranging from pick-your-own to modest roadside stands to elaborate and well-maintained attractive retail outlets. Some respondents, particularly those selling livestock and eggs, undertook no special measures; many sales on such farms were transacted in the farmyard. In several businesses, however, farm produce was sold in conjunction with produce from other farms, arts and crafts, and baked goods.

Table 4.4 Farm operator adaptive behaviours near Toronto

Adaptation	Number of farmers reporting	Percentage of total
Cash cropping on rented land	60	57.1
Direct-to-public sales (including Pick-Your-Own)	31	29.5
Horses – livery and/or breeding	14	13.3
Total	105	100.0

Source: Johnston (1989)

These patterns are fairly common. Some direct comparisons between the Toronto and Montreal areas (Table 4.5) show the similarity of the movement towards greater diversification over the study periods. The somewhat greater diversification of the Montreal farms in marketing strategies reflects the much lower level of this activity 'initially', compared with the Toronto area.

Most of the respondents engaged in this adaptive strategy depended upon the non-farm population to support their endeavours, drawing most of their clientele from Toronto. Several of the more entrepreneurial suggested that a location about forty minutes' driving time from downtown Toronto was just about ideal to tap that market. Forty minutes was a journey time long enough to generate a 'sense of occasion' for many urban families, but not long enough to induce weariness. These sorts of concerns are consistent with Laureau's (1983) analysis of pick-your-own customers north-west of Paris.

A reverse process of extensification can also accompany farm business expansion, primarily when there has also been expansion of the physical land of the farm. This is therefore linked to the general process of farm consolidation and amalgamation, specialisation and capital substitution for labour, all part and parcel of the quest to achieve economies of scale in production. In the inner urban fringe, this has often been reflected by cash

Table 4.5 Farm diversification in the Montreal south bank (1981–88) and Toronto SW (1975–85) areas

Zone	% farms reporting being engaged in	
	Farm enterprise diversification[a]	Marketing strategy diversification[d]
Toronto[b]		
All	41	19
Zone I (most urban)	42	25
Zone II (semi-urban)	52	25
Zone III (rural)	26	4
Montreal[c]		
All	42	28
Zone I (most urban)	47	37
Zone II (semi-urban)	44	28
Zone III (rural)	29	21

Note

(a) Defined in terms of net increases in number of farm enterprises over the respective study periods
(b) Based on 189 farms
(c) Based on 149 farms
(d) PYO, value added, recreational enterprises

Source: Abstracted and compiled from Marois et al. (1991)

cropping on extensive tracts of rented land. However insecure such operations might appear to be, there is evidence that this can remain a fairly permanent adaptation, necessitating the relinquishing of land parcels in one location only to pick up others somewhere else. In Johnston's sample near Toronto (1989), 60 of the 189 farmers surveyed were involved in cash cropping on rented farmland. In that study, this adaptive strategy was defined as involving only those farmers who had acquired rented land solely for the production of cash crops. The most typical crops produced in the Toronto area were corn, wheat and soy beans, depending mainly on market conditions. In most cases, cropping augmented an existing livestock operation, but some respondents had taken the opportunity to reshape their business with cropping becoming the dominant enterprise.

Cash cropping is heavily dependent upon capital, but much of the money invested is in mobile assets like machinery, particularly when rented farmland is involved. It is an attractive option for those farmers who enjoy working with machines or who have become tired of continually looking after livestock. On the landscape in the urban fringes of south-west Ontario, cash cropping can be identified by the large, round metal grain storage bins and the confusing array of long tubes which connect the bins in no apparent order.

The availability of large tracts of rental land was an important condition for this adaptive behaviour. This may be a matter of concern since many commentators contend that rented farmland is neither managed nor maintained as well as owned land. The commonly expressed fear is that many farmers 'mine' their rented land and that soil structure and fertility suffer as a result. However, as reported by Johnston and Smit (1985), empirical evidence can be cited which both supports and undermines this opinion. Indeed it can be argued that the problem of land mining has as much to do with the nature and stability of the lease agreements themselves as with the fact of renting.

Some respondents engaging in this adaptive behaviour in Johnston's study did admit to 'mining' some of their rented parcels. Upon closer questioning, though, it was discovered that such practices typically occurred when rental leases were short or unstable, or when development was imminent. Most respondents, especially those who had secured written leases, reported using their rented land in the same way they used their owned land. Most of those interviewed considered renting farmland to be a viable alternative to purchasing land, particularly in light of very high land values in parts of the study area, and few of those engaging in this adaptive strategy voiced any serious fears or apprehensions regarding the practice. This is consistent with evidence that Munton has discussed for farmers in London's Metropolitan Green Belt (1983a) as well as Bryant's analysis of French farmers around Paris (1981).

A final form of expansion of farm business is through the addition of non-traditional enterprises on the farm unit. In addition to commercial outlets such as garden centres and the like, these non-traditional enterprises include non-agricultural enterprises such as the storage of recreational vehicles and the provision of parking facilities for a nearby urban clientele (Bryant, 1981) as well as non-traditional agricultural enterprises such as livery stables and stud-farms. Gilg (1990) calls this type of adjustment a 'structural' form of diversification, where the addition of non-agricultural enterprises none the less utilise 'agricultural' resources on the farm. Of course, it is debatable whether some of these enterprises are entirely non-agricultural, since some of these non-traditional activities such as stud-farms and livery stables do have ties with agricultural activity.

In Johnston's study, fourteen of the respondents surveyed indicated that they had made a decision to set up a livery stable and/or breed horses because they were located in an urban-centred region. Two very different kinds of farm operations fell into this category. One group was comprised of people who had previously been livestock farmers (mostly dairy) and who were now semi-retired. Using their farm buildings and land to keep livery horses for non-farm customers provided them with a source of income. The work required to run such an operation is not as demanding as dairy farming and in the words of one respondent: 'It allows me to keep in farming without all the work and worries of milking cows.'

The other group engaged in this adaptive strategy were part of the horse-racing industry, and their operations were more highly commercial. In several cases, farms were owned by a group of business partners who became involved in breeding racehorses as an investment. For these businesses, their location decision was strongly influenced by proximity to Toronto, which in turn is the reason that several horse-racing tracks are located in or near the study area.

In other areas, riding-schools exist either alone or in combination with livery stables and stud-farms. Sometimes they are non-traditional enterprises added to a farm; but frequently, they are specialised businesses whose operators have not had any 'traditional' farming experience. This was the case, for example, in Benmoussa et al.'s (1984) survey of horse-riding establishments in the Chevreuse valley south of Paris.

Some farmers make the decision to contract the farm business (and we do not include the forced decisions resulting from, say, expropriation of land for non-farm development purposes or the taking back of rented land by the owner for some purpose). Contraction of farm business may involve extensification, with the shift to more extensive and less labour-intensive and/or less capital-intensive production techniques. It can also include reduction in the acreage by selling part of the land either as a severance or for some other purpose. As with the expansion strategies, many factors can be expected to influence such decisions, including personal and family circumstances, local development pressures, the financial state of the farm business and so forth.

Sometimes, the contraction may be related to other policy options, namely the decision to seek off-farm employment or to retire or seek another full-time occupation (Figure 4.5). Seeking part-time employment, either on the part of the farmer or members of the farm family, may lead to a reduction in the amount of family labour available for farm work compared with the quantity already used. In this case, simplification of the enterprise structure in order to make fewer demands on the labour supply may well occur, even to the point of setting aside some of the land. Of course, where the farmer and family labour transferred to an off-farm occupation are surplus to the farm's current needs, then no such adjustment is necessary. Where an adjustment is made, it is appropriate to see this as a move to a new farm structure, rather than another stage in the transition of the farmer out of agriculture (see Chapter 3 for a discussion on this).

In the case of retirement, winding-down strategies are more likely to be encountered, as the farmer decides to reduce his or her level of activity progressively. Sometimes this is precipitated by ill-health, and sometimes by a son or daughter deciding to leave the farm. In addition to these factors, farmers (see Chapter 3) may decide to wind down the level of activity in anticipation of urban expansion (Bryant, 1974; Wibberley, 1959), disinvesting in the land resource and 'mining' the soil. A complete occupational shift out of agriculture is more likely to be accompanied by sale and liquidation of the whole farm, although in some situations, the former farmer and farm family may retain ownership of the land and rent it out to other farmers.

Finally, cutting across all of these strategies are various organisational strategies, which may be taken in order to pursue the policy option of restructuring farm debt and capital ownership. This can include the development of corporate farm ownership to facilitate successoral continuity of the farm in the farm family as well as extended forms of family farm business structure. As a result of these types of decisions, it may sometimes appear that owner-occupiership of the farmland has increased (e.g. in Munton et al.'s sample (1988)); however, this may hide shared family ownership of farmland and the farm business as well as various forms of profit-sharing arrangements between the farmer and family on the one hand and other 'sleeping' partners in the farm on the other, either family or non-family members. Such arrangements are not always entered into freely by the farmer. For instance, in some of Bryant's fieldwork in the Paris region, France, a number of cases were found of shared family ownership of the land following the death of the parents. The shared family ownership was simply a solution to deferral of estate duties but which necessitated the farmer paying rent to his or her siblings and sometimes involving them in profit-sharing too.

The whole issue of trying to measure the independence, and security, of the farm family by looking solely at owner-occupiership is complicated. Other forms of external control over the assets of the farm business also exist, such as indebtedness to commercial banking institutions. High levels of owner-occupiership may give rather an illusory image of independence (Johnston, 1990; Munton et al., 1988).

Factors influencing the choice of strategies

Farmers differ in the strategies they adopt to cope with and exploit the conditions which prevail in near-urban zones. This, of course, is why we observe variations in the structure of agriculture near cities, or anywhere for that matter. Farmers' choices of policies and strategies in the City's Countryside are affected by a wide range of interrelated factors.

Structural conditions of the broader environment can help shape adaptation in the City's Countryside, and some of the changes in this environment, as already noted earlier, are related to the development of post-industrial society. General broad-based changes in consumer preferences towards healthier, fresher food has played a major role in generating opportunities for some adaptive strategies. Of course, there is no guarantee that farm-purchased products are necessarily any healthier (e.g. free of pesticide residue) than the same item purchased from a supermarket. Still it is what the consumer believes that matters in the final analysis.

Some farmers may also be forced by structural circumstances to engage in adaptive strategies, particularly if those strategies improve the economic viability of the business. The high cost of borrowed money, particularly in the last quarter of the twentieth century, combined with large debt loads has meant that farmers in the City's Countryside and elsewhere have had to restructure their operations in order to improve cash flow or increase revenues or both.

Other characteristics of the macro environment can present conditions favourable to certain types of strategies. For instance, on the positive side, the support offered to pick-your-own operators by the Ontario Ministry of

Agriculture and Food in the form of production and marketing advice and advertising through a Ministry-developed guide to pick-your-own growers has undoubtedly helped to create opportunities for new growers (e.g. Ontario Ministry of Agriculture and Food, 1984).

Location can also play an important role in a farmer's decision to engage in an adaptive strategy, particularly in view of the fact that most farmers start off with their location, and therefore the characteristics of the land and soil as well as its situation are already determined. The problem, therefore, is to select an activity which conforms to both the site and situational characteristics of each farm.

We can therefore think of different locations offering particular opportunities (i.e. possible strategies), while at the same time imposing constraints on choices. Constraints and opportunities exist as a result of a complex interplay of a particular site's physical capabilities and its location *vis-à-vis* various systems of exchange, as well as its relationship with the macro and micro-enabling environment. For instance, planning regulations and policies of various kinds can create some opportunities while constraining others. Land use planning regulations, which restrict the types of non-farm activities in a particular zone that can be developed by farmers on their farms, may in the end seriously undermine the economic viability of the farm family.

Other factors relate more to the economic and socio-economic characteristics of the farm and farm family respectively. In terms of the individual farm, the production systems present on the farm when it is decided to make an adjustment can also help shape the nature of the strategy adopted. Most dairy farms in Southern Ontario, for example, produce crops for cash, so the expansion of this part of the business can be done relatively easily. Beef producers typically meet their household's needs for beef, so it is not much of a departure to start supplying other households. Similarly, farm gate sales for most fruit and vegetable-growers is very much a logical extension of the existing enterprise. On the other hand, expanding dairy production in areas with supply management in place is a much more difficult proposition.

In terms of factors that relate to the farm family and farmer, one of the most important subsets is their goals and objectives (see the earlier discussion of goals and values in this chapter). Most of those farmers undertaking adaptive strategies in the Toronto fringe study, for example, were highly committed to farming for both occupational and life-style reasons. Most were also deeply attached to their geographic area; many, for example, had family roots in the region dating back several generations. Another factor that figured prominently in decisions to undertake adaptive strategies was whether or not the farmer's offspring were interested in taking over the family farm. Indeed, Johnston (1989) found that while 27 per cent of his total sample of farmers said that their children would be taking over the farm, 37 per cent of those undertaking one or more of the adaptive strategies fell into this group. Conversely, farmers who reported that their children had no interest in taking over the family farm were much less involved in adaptive strategies.

Focusing on the individual farm level emphasises the variability of responses related to differences in the farm structures and the farmer and farm family characteristics. An interesting question that arises is whether there is any systematic variation in this variability itself. Bryant (1981) hypothesised that if there is any systematic variation in the stresses or stimuli influencing the farm decision-making, then it would be reasonable to expect that the range of variation of farmer evaluations of those stresses and their reactions to those stresses would vary too. A high degree of variation could be expected where the stresses were ambiguous, because this would allow for a greater variety of factors to influence the evaluations. Conversely, where the stresses were objectively very high, the evaluations would be more likely to cluster.

Bryant developed this idea in relation to the evaluation of the likelihood of future urban expansion by farmers; the framework is summarised in Figure 4.6. In this framework, the 'average' expectation of a non-agricultural future for the farmer's land is suggested as being directly related to various indicators of potential urban development, e.g. proximity to urban development zones, prior experience with selling off land for development purposes, and so forth. However, it was also hypothesised that the range of variation of farmers' evaluations would also be related to the actual strength of these indicators, being relatively narrow in areas where the strength of urban development indicators was either very strong (e.g. the inner urban fringe) or very weak (e.g. the outer parts of the urban shadow zone). Only in the intermediate situation, where there was more 'uncertainty' in the strict sense of the term, was a significant range of variation in farmers' evaluations expected, their evaluations being influenced by many other factors when the urban development potential indicators were ambiguous.

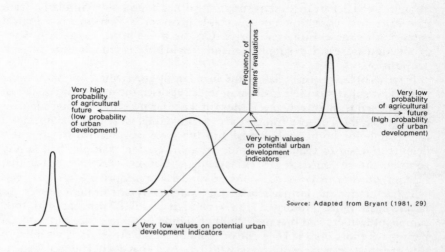

Figure 4.6 Farmers' evaluations of an agricultural future for their land and potential urban development

In Bryant's sample of farm interviews, farmer evaluations did indeed conform to the hypothesised pattern. However, there was no systematic relationship between these evaluations and actual farm investment behaviour (i.e. the farm strategies adopted). Bryant (1981) suggested that this was due partly to the other positive aspects of an urban fringe location such as direct marketing opportunities even in the areas with a high potential for urban development and partly to the existence of the many other factors that influence farm change at the individual farm level. The fact that there is quite a variety of strategies that can be adopted to address the same type of potential urban development, even if different farmers evaluate the pressures similarly, renders the situation even more complex.

Evolving farming landscapes in the City's Countryside

How do the various forces of change in farming in the City's Countryside combine to influence the geography of farming there? The perspective taken here is to integrate some of the previous frameworks introduced, notably those represented in Figures 1.3 and 4.3, and suggest that there is an emerging pattern. This pattern is not so much represented by any zonal geographic structure, but by a mosaic or complex layering of farming structures whose variability none the less may vary in some systematic manner from place to place within the City's Countryside (Ilbery, 1988a).

Bryant (1984a) in fact suggested that the metropolitan forces and non-metropolitan forces of change as well as the regional (and local) environment influences combine to produce different types of farming landscape and structural change. The framework builds directly out of that developed for Figure 1.3, and recognises on the one hand both positive and negative influences of any or all of the forces with respect to agricultural productivity and, on the other hand, the importance of adaptive, entrepreneurial behaviour. Landscapes of agricultural degeneration are dominated by negative forces. For instance, urban development pressures, the indirect impacts of non-farm development upon farm structure and poor farm structure exacerbated by intense competition for the agricultural specialisations of an area may combine to produce environments that are very difficult for farming. Perfect examples are found in some of the old-established intensive agricultural zones around some of the West European cities, such as Paris (Biancale, 1982; Bryant, 1984a).

Landscapes of agricultural adaptation, it was suggested, also suffer some of the negative impacts of urban and non-farm development pressures and perhaps some inherited problems of poor farm structure; however, they are not as severe as in the first type of landscape. Indeed, these negative effects can be outweighed by various positive influences of other factors such as direct marketing opportunities, excellent soil and readily available land for rental. Some farmers are able to adapt to the pressures and take advantage of the positive influences, while still others may become involved in more proactive entrepreneurial activities. Finally, landscapes of agricultural development are suggested as those areas where no major urban development pressures or other negative influences are present; the evolution of the farm structure and landscape is primarily the result of non-metropolitan factors as well the metropolitan labour effect. Both the landscapes of agricultural adaptation and development involve complementary relationships between one or more of the urban-based forces and agricultural development (Bryant, 1984a, b).

We should be careful not to draw the same erroneous conclusions regarding level of development pressure and landscapes of agricultural degeneration which were made in studies on farm/non-farm population ratios (see Chapter 1). Indeed, in Johnston's (1989) study, while his most urbanised zone certainly contained landscapes of agricultural degeneration, it was also the zone in which farmers expanded their acreage as frequently as in the most rural zone, and more farms expanded their labour inputs than elsewhere (Johnston and Bryant, 1989). Similarly, Deslauriers et al. (1990) show that their most urban zone had a higher proportion of expanding farms (1981–8) than the rural zone, as well as a greater concentration of specialised farms (e.g. horticulture, vegetables).

It should be clear that this framework does not suggest any sort of homogeneous zonal structure; indeed, the evolution of the structures and landscapes takes place through individual farm level changes. Apart from personal differences between farmers and farm families and differences resulting from local variations in the farms' biophysical environments, interfarm variation is partly to be explained by the variety of farming methods and by the different systems of exchanges in which farms can be involved. Therefore, any systematic variation in the geography of these evolving structures and landscapes, which is certainly suggested, is really a statistical generalisation of tendencies and sometimes quite disparate individual farm level behaviour. This complexity can perhaps best be represented as a complex mosaic of structures and changes (Bryant, 1989c). The reality of this mosaic is evident in study after study (e.g. van den Berg's and Ijkelnenstam's (1983) analysis of part of the Green Heart of Randstad Holland), even though we have tended not to dwell on this phenomenon.

The City's Countryside: new opportunities for farmers and farming

The City's Countryside is a dynamic environment for farmers and farming. Farmers in near-urban zones have long been at the forefront of change and there is no reason to think that the future will be any different. Before moving on to discuss public intervention in agriculture in the City's Countryside, a brief set of summary comments are made concerning what appear to be emerging patterns.

The City's Countryside will probably be the site of continuing tension between groups holding very different sets of values. On the one hand, structural conditions affecting agriculture will undoubtedly continue to foster the adoption of industrial modes of production, with all that that means in terms of manufactured and chemical inputs, large-scale machinery, removal of fence-rows and so on, while increasing pressure will be brought to bear by consumer groups, rural non-farm residents and conservation interests for the adoption of ecological farming practices. It is not clear at this early point in the debate how much room for compromise exists.

Perhaps part of the answer lies in abandoning agricultural production goals as the primary goals for the City's Countryside. In the European Community with its massive problems of agricultural overproduction and the huge public costs of agricultural support, this transition has already occurred to all intents and purposes in several regions. Our interest then focuses on the degree to which agriculture provides a means of maintaining open space and an attractive landscape, a model of which has already been implemented in parts of Western Europe, including the National Landscape Parks in the Netherlands. Embodied in the concept of these Parks is the

expectation that landholders, in addition to working their land, should assist in the management of the park and receive payment for activities concerned with its care, as well as compensation for loss of income as a result of any limitations on land use. Thus farmers, for example, would no longer merely supply agricultural produce, but also provide the community with an attractive landscape. (Pigram, 1987, 66)

However, it is highly doubtful that this kind of scheme could be introduced in the near future in places like Canada or New Zealand, where the ethic of private landownership is simply too deeply entrenched. After all, the ancestors of many of these countries' modern farmers left their homelands so that they could become landowners themselves. We can hold little hope that these people will rush to endorse a proposal that would act to constrain their rights as landowners. Another reason for countries outside Europe being likely to resist the Netherlands' model is that there still exists in these places a strong bias toward large-scale production agriculture. Many farmers, for example, still judge themselves and their neighbours by the size of their tractor. This kind of perspective has seeped into the farming subculture over many generations and it will not easily be reversed.

Nevertheless, farmers may well be forced to change their methods in the coming decades. An increasing number of commentators pointing to such problems such as high energy use in agriculture (Lockeretz, 1981), the harmful health effects of modern production systems, both on the farmers themselves and the public at large (Coye, 1986), and soil erosion (Pierce, 1987) have cast serious doubt on the sustainability of modern production systems. Pressure is already being brought to bear on agricultural interests to develop sustainable production systems, and the City's Countryside will probably be one of the major battle grounds. This is, of course, because it is in near-urban zones that the interaction between agricultural and other land-use interests is at its most intense.

The future could also bring with it a shift in focus for farmland conservation policies. Schemes which aim at the preservation of the land resource have not worked particularly well (see Chapter 5): the least effective strategies have been widely implemented and the most effective strategies hardly at all. Part of the problem has been the lack of attention given to the individual farmer, farm family and farm business (Pyle, 1989). Clearly, farming is, at its roots, an economic activity, and unless it can be pursued in such a way that it affords farm families an adequate life-style, then it does not really matter how much farmland we have managed to save from being covered with bricks and mortar. If agriculture is to be retained in the City's Countryside, whether the goal is production or not, we need to recognise that people do the farming and the farm business is, and will likely remain, the basic unit of production. Where we are dealing with farms for which economic goals of agricultural production are not uppermost, it seems almost perverse that some of our public intervention strategies have sought to exclude them from agricultural zones in the City's Countryside even though they may represent a very stable element in land use.

We have stressed in this chapter how important it is to understand the farm as an operating production system, combining people, land and capital. In much of the agriculture in the City's Countryside in the developed world, various forms of the family farm dominate, with its special combination of the economic business unit and the family unit. The goals and objectives of these decision-taking units have to be considered by any public sector attempt to intervene in the processes of change, partly because as individuals and families the farm population must be taken into account in any democratic system, and partly because in a non-totalitarian system, the actions of individuals can easily thwart the achievement of broader collective goals. The challenge therefore is how to integrate the collective goals that we might define as a society for the human and physical resources involved in the farming landscapes in the City's Countryside with the individual level of the farm and farm family, and the need to recognise the heterogeneity of individuals and socio-economic structures of production and life-styles.

5
The government: intervener in the enabling environment

Introduction

There are many different goals and perspectives for the human and physical resources involved in agriculture in the City's Countryside. In Chapter 4, stress was placed on the perspective of the individual farmer and the farm family. Already in the conclusions to the last chapter, the significance of this was suggested, even in terms of planning and managing the farm system in order to achieve broader collective or societal goals. As Pyle (1989) points out, it is really quite naïve to expect landowners (and farmers) to make the same decisions just because they happen to be in the same area.

The public sector through the various levels of government plays a major role in shaping the environment within which farming operates, by altering the rules, regulations and/or values attached to various systems of exchange. Public sector involvement in agriculture, for instance, can include the approval and inspection of chemical pesticides and fertilisers or the granting of permission for land development on agricultural land. It can also include intervention that is more facilitating or enabling such as through the creation in some countries of incentives for farmers to join together in soil and water conservation associations or influencing the cost of borrowing money for capital investment on farms.

In relation to any particular component of the farming system – land, people, capital – public intervention is complicated because the component can be influenced either directly or indirectly. For instance, public intervention can affect the patterns and stability of farmland in the City's Countryside either directly through land use policy or indirectly through price support policy, income stabilisation schemes and fiscal, tax and trade policies.

What complicates the matter even more is that frequently these indirect influences of public intervention are unintended. In this sense, public intervention can become simply another set of factors with which farming, the farmer and family have to contend. The widespread movement towards 'deregulation' and generally less government intervention can also be seen in this light. Governments have frequently intervened in agriculture through subsidies, price support and the like. When attempts are made to remove or reduce such intervention, for whatever reason, stresses are placed on farmers' operating environments modifying the systems of exchange in which they function.

Logically, public intervention in farming can be seen as responding to a set of defined goals, both involving the farm sector itself and its relationships with the non-farm segments of the socio-economic system. The problem, of course, is that we are dealing with a complex system, already undergoing complex patterns of evolution without public sector intervention. In these circumstances, it is unlikely that intervention in one component of the system can be undertaken without affecting other components of the system. Furthermore, there is no reason to suppose that all goals defined for public intervention in the agricultural sector are automatically compatible with each other, and this is certainly the case when we consider policies for the non-agricultural sectors of the economy as well.

A final complication is that public intervention which may have an impact upon agriculture can be developed and implemented on different levels reflecting the different levels of government in any given country (local, region/county, state/province, country, and even international). Furthermore, the degree of centralised versus decentralised responsibility for certain types of public intervention varies significantly from one country to another, and even within countries, especially where there is a federal structure (Bryant and Russwurm, 1982).

Finally, the organisation of intervention within government differs enormously, ranging from extremely sectorally defined organisations to more integrated mechanisms. What this means is that agriculture in the City's Countryside may be simultaneously responding to public sector intervention conceived to deal with particular issues in the whole agricultural system or country, dealing with the unanticipated side-effects of public intervention, and also in a more positive light responding to intervention designed to help it cope with challenges that are particular to the City's Countryside. Our focus is on the last category, but it is well to remember that the others exist too.

In this chapter, some of the reasons and approaches taken by the public sector to farming and farmland in the City's Countryside are discussed. Farmland is used as the principal theme of the discussion. This is not because less importance is attached to people and capital in farming, but simply because there is so much more research and public sector activity aimed at farmland in the City's Countryside. The discussion highlights, however, the importance of human questions in the development of public sector policy and interventions in the coming years. The question of the relationship between intervention in farming and broader issues involved in sustainable development are also examined.

First, we consider the range of rights and interests in farming and the various actors who become involved in intervening in the transformation of agriculture and agricultural areas. Second, the general rationale for public sector intervention in the systems of exchange in the City's Countryside is outlined and the challenges discussed. Third, a general framework for considering the different approaches and techniques that have been and could be taken regarding agricultural land in the City's Countryside is presented. The types of approaches and techniques used in farmland conservation schemes are dealt with in some detail as they represent the major set of interventions aimed specifically at agriculture in the City's Countryside; the issues involved in the evaluation of public intervention, using farmland conservation as the major example, are also examined. Then we consider explicitly the other important dimensions of agriculture in the City's Countryside. Finally, the discussion of goals, interventions and the range of agricultural situations is synthesised at the end of the chapter.

Rights and interests in land

Interests in land

While the focus in the discussion on rights is on land, the discussion of the broader interests in farming must of necessity give some consideration to people and capital. In Chapter 1 a framework was used to identify the broad range of functions that agricultural land can be called upon to perform. That framework is repeated here for the sake of convenience, with some additional detail (Table 5.1).

The broad functions of agricultural land are production (agricultural production), protection, place and play functions. Both private and public or collective interests are embedded in each of these sets of functions. Most of the private (or individual) interests in land can be established through purchase of rights in land (see below) and the market system confers a value on the land based on the purchaser's willingness to pay. This in turn is linked either to the individual's anticipated enjoyment of the land (e.g. hobby farming, residential development) or to how much a user (as opposed to owner) of the land is willing to pay for the product or service produced or supported by the land (e.g. the price of agricultural produce, user fees for livery horses, user fees for other recreational enterprises on the farm). Some private interests in agricultural land do not, however, involve user fees, e.g. the public use of private agricultural land for recreational activities such as cross-country skiing and hiking. Some public or collective interests or values in agricultural land can also be expressed through the market with a public agency purchasing rights in land and establishing a proprietary interest in it, but they are mostly pursued through some form of regulation of the use of land. This is especially the case for the collective interests in the production functions of the land.

Table 5.1 Private and public interests in agricultural land

Broad functions	Private interests	Collective interests
PRODUCTION	Farm level production – food – non-food products – services	Potential food production Access to food supply Support of food processing and agricultural supply industries Maintenance of a farm population and community Food aid Balance of payments
PROTECTION	Private reserves	Wildlife habitat support Water supply Flood plain management Soil conservation (erosion, degradation) Conserving open space to control urban structure
PLACE	Housing Industrial and commercial development Agricultural production oriented to specific markets Recreational enterprises	Access to housing Support for economic development and employment Infrastructure support
PLAY	Hobby farming Recreational enterprises on farms Public use of private (farmland) lands	Amenity value of agricultural landscapes Providing access to recreational opportunities for the population

Many of the collective interests in agricultural land represent a concern for people (Table 5.1). From the perspective of the sustainability of food production systems, people issues are really the primary concerns. The food sufficiency issue is directly related to concern over the capability of the agricultural system with all its resources to feed the population. Collective interests in the production capacity of the agricultural system linked with food-processing and agricultural supply industries, as well as its contribution to foreign earnings all essentially reflect concern for people – their incomes and their quality of life.

Collective concern related to maintenance of a viable farm community and population base are clearly linked to concern for people too. In relation to public sector intervention in agriculture, this concern over people extends beyond the City's Countryside, although we can expect issues such as the parity of income between farm and non-farm sectors to be more acute in the City's Countryside.

It is important to keep in mind the human dimension of agricultural land. It implies that public sector intervention that is directed towards the land or its particular elements and indeed the capital base of the farm is undertaken in the long term for the sake of people, either farm or non-farm. Furthermore, the link between agricultural land, people and capital implies that public sector intervention explicitly to address, say, low incomes in farming or the actual or potential failure of farm incomes in an area resulting from poor market or weather conditions also has significant implications for agricultural land – its continuance in farming, how and for what purpose it is used.

When the discussion is broadened to include collective interests in agricultural land because of its potential to support non-agricultural functions, the human dimension is clearly paramount. On the one hand, some of the collective interests in agricultural land as a support for other functions are incompatible in the long term with continued agricultural production because conversion of land use is the result. None the less, a legitimate collective interest can be established in managing a particular area of land in order to maintain its potential for some other use. On the other hand, other collective interests can be realised without removing land from agricultural production, e.g. the management of a flood plain area or having agricultural land used simultaneously for recreational activities.

Up to this point in the discussion, collective interests in agricultural land have been simply set against private interests. The distinction is important because conflicts arise between private and collective interests which need to be resolved. However, collective interests do not represent one monolithic set of interests. First, conflicts frequently exist between the multiple collective interests, e.g. between interests in maintaining food production potential and using the land for housing or economic development in the non-farm sectors. Second, collective interests exist on a variety of levels: local, regional, national, international. There is a certain correspondence between these 'levels of interest' and the hierarchy of public sector involvement: local municipality, region or county, state or province, central state, supra-national grouping. This adds considerable complexity to our analysis because some of the conflicts between collective interests in agricultural land reflect different constituencies of interest. For example, the debate between the use of agricultural land for food production as opposed to housing, industrial and commercial development is one that generally pits a broader collective interest against a more local interest. The increasing significance of local level public sector involvement in planning and management of the land resource in many jurisdictions poses some difficult challenges in integrating all of the collective interests into the way we deal with land. We shall return to this later.

In the developed nations of the world, both private and collective interests have to deal with land ownership or the rights that exist in the land. Because of its significance, land ownership is discussed next.

The ownership of land

The right to 'own', use, develop, sell, lease and mortgage property is one of the fundamental cornerstones of Western civilisation and is particularly well entrenched in the New World of North America, Australia and New Zealand. The ownership of land and the operation of markets in land are complex topics which have produced an extensive literature (e.g. Barlowe, 1986; Bryant, 1972; Goodchild and Munton, 1985). In the following discussion, only a few major points are made to set the scene for the analysis of public intervention in the land market in the City's Countryside which affects agriculture.

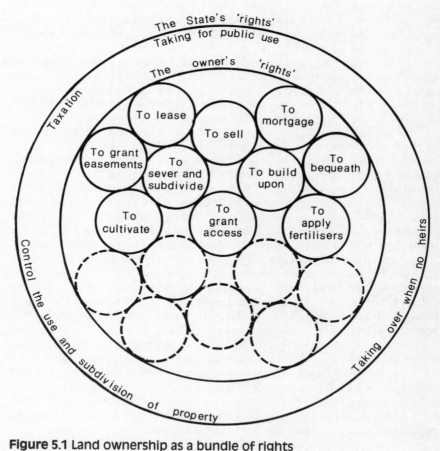

Figure 5.1 Land ownership as a bundle of rights

One of the most important points is that there is a significant difference between the lay person's understanding of property ownership and its legal meaning. Lay people continually have difficulty with the notion that ownership of the land, while conferring on them exclusive rights, does not give them absolute rights over the use of that property. It is useful to think of land ownership as a 'bundle' of rights, which exists within a macro environment in which the State can exercise certain rights and a certain degree of control over private rights (Figure 5.1) (Barlow, 1986; Bryant, 1972; Denman and Prodano, 1972).

Private rights in land include the right to use the land for personal enjoyment and for agricultural production, the right to develop it, to sell it, to lease it, mortgage it and to subdivide it and the right of access to the land. These rights can be detailed even further to include such things as the right to erect bill-boards and other signs, the right to alter structures erected on the property and so on.

There are three important points to note about this 'bundle' of rights concept. First, given that the different rights can be identified and separated from each other, a particular right can potentially be sold, leased or otherwise exchanged without affecting the property owner's continued enjoyment of the other residual rights. Second, following the same line of argument, public sector intervention can be thought of as being aimed at particular subsets of the whole range of rights existing in a property. Third, the restrictions on rights in a given property can affect the market value of the land either because some rights have been traded away, temporarily or permanently, or because of public intervention, e.g. restrictions on the uses to which a property can be put in a land use plan or restrictions on the degree to which a property can be subdivided. Indeed, it is more proper to speak of the value 'in' land in order to recognise that it is a specific bundle of rights associated with a given property that is being bought, not the land itself (Denman and Prodano, 1972).

Governments are therefore able to restrict the bundle of individual rights in land by various means. These collective rights are generally exercised in order to protect public health and safety, which can be defined in a very generous fashion, or to achieve some other objective determined to be in the public interest. Depending upon the jurisdiction, these restrictions are exercised and policed by different levels of government, e.g. local municipality, region, county, state, province, central state government. From the perspective of agriculture in the City's Countryside, the most frequently used restrictions that affect agricultural land are those which deal with land use, e.g. land use plans, zoning by-laws and their accompanying regulations, and property subdivision. In addition, governments can exercise other rights (Figure 5.1) to raise revenue (right to tax property, right to take and sell the property of a deceased person without any heirs) and to take over the land to provide for some public need.

Attempts by governments to exercise their right to control private property owners' use of their own land can result in heated and protracted debate, partly because of the difficulties often involved in defining the public or collective interest and partly because of the strong attachment to private rights in land in many countries. This is particularly the case in the United States where, compared with countries such as Great Britain and Canada, it has been much more difficult for governments to place limits on the use of private property.

Rights and interests in property and the land resource are dynamic (Grove-Hills, 1990). New interests associated with property are evolving continually and they can quickly become linked to new rights in land which can be traded in the market or restricted for the public or collective benefit (Braden, 1982). An excellent agricultural example is the increasing public and public sector interest in conservation and sustainable development (Munton, 1983b; Phillips, 1985; Swain and Haigh, 1985). Public sector intervention, e.g. encouragement to farmers to adopt more conservation-oriented cultivation practices, heralds the establishment of a new public interest in agricultural land. When that interest becomes expressed as policed restrictions on the type and quantity of chemical fertilisers, pesticides and herbicides that can be used, then it is also clear that we are talking about a restriction on another right.

Some of the resistance to public control of land use is ideological. However, much of it also stems from the opportunity costs that public control and intervention generate for private landowners. The following example illustrates the types of costs that can be involved for a landowner. The example is one of many cases where public sector intervention has sought to protect prime quality agricultural land from urban development by proscribing urban development from certain areas. Based on the above discussion, an expected consequence of this type of intervention should be a reduction in the market value of the land because of the restriction in the range of rights. New Zealand Valuation Records show that good quality farmland on the periphery of Palmerston North, a city of nearly 70,000 located 145 kilometres (90 miles) north of Wellington, New Zealand's capital, was sold for farming purposes in late 1989 for $4,200 per hectare. Similar land at the urban margin sold for residential development commanded a price of $29,166 per hectare. Given such large differences in land values depending upon the market in which it is sold, it is hardly surprising that farmers often hold the opinion that they should be 'free' to dispose of their land as they please. For example, 71 per cent of a sample of 310 farmers from the fringe of Brisbane in Australia held the opinion that private landowners should be free to sell their land to the highest bidder, regardless of its future use (Rickson and Neumann, 1984).

Actors involved in agricultural land issues

Consideration of different interests leads naturally to considering the different actors involved in the response to the issues associated with agricultural land. The specific set of actors varies according to the particular issue, e.g. farmland conservation versus soil conservation versus landscape conservation. Taking farmland conservation as an example, four distinct groups of actors have each played a role in the development of farmland conservation policy. Some play a direct role (e.g. planners), while others play an indirect one (e.g. the public through their voting patterns, or through various interest or lobby groups). Furthermore, some groups of actors (e.g. farmers) are directly affected by farmland conservation, others much less so.

First, of course, *farmers* appear to have the largest stake in the farmland conservation debate. If their land is ripe for development and it has been frozen in agricultural use, then we have already seen that they can experience substantial opportunity costs. On the other hand, if a farmer is keenly interested in remaining in production in a rapidly urbanising area, then government action may well serve that interest.

Researchers have found farmers' views on the subject of farmland conservation policy to be complex and, at least superficially, contradictory. For instance, while many farmers agree with the general principle that good quality farmland ought to be retained for agriculture, it is clear that many farmers are more than willing to sell their properties even if they know it will be developed. Unless they are in a situation where a farmland protection scheme will impose an opportunity cost on them, most farmers will support the scheme. This simply reflects the fact that most farmers wear two hats – as a farmer for whom the land is a means of making a livelihood, and as a landowner, one of whose main interests is in maintaining and capitalising on the value of his or her assets in land. In an industry where low returns have plagued producers, it is easy to feel sympathetic for the plight of farmers who look on their land as their security and pension fund.

It is not surprising therefore to find many farm groups opposed to farmland protection policy, at least initially (Detwiler, 1980; Macpherson, 1979). In North America, this is also related to the strong local or 'bottom up' emphasis in planning which in turn is reflected in the voluntary nature of farmer participation in many farmland protection programmes such as the Williamson Act of California (Davies, 1972), the New York Agricultural Districting law (Sullivan, 1977) and the Wisconsin program (NALS, 1980). Of course, this does not mean that there are not some farm groups who are very supportive in the development of farmland protection policies and some individual farm leaders who have had considerable influence (e.g. Little, 1974).

Secondly, there are some *politicians* who support farmland conservation policy for ideological reasons, as indeed there will be some politicians who oppose farmland conservation because of a philosophical aversion to government intervention in the economy. However, politicians have a strong tendency to take their lead from the voters, on this as on many other issues. This is one of the characteristics of the democratic systems which dominate developed countries. Research has shown that attitudes held by the general public can exert a powerful influence on the adoption of farmland conservation policy. For example, Furuseth (1985a, 1985b) found that counties in California with established liberal political traditions were much more likely to have implemented robust farmland conservation measures. The reason for this is undoubtedly that liberal Californians accept public intervention more readily than their more conservative counterparts.

The *public*, a rather amorphous label, constitutes the third category of actors. It is however a heterogeneous category, ranging from John and Jane Doe who may have a very poorly articulated perspective regarding agriculture and agricultural land to special interest groups with strongly held positions and well-articulated perspectives. In terms of the general public, it is interesting to note that non-farmers have often played an important role in the development of farmland conservation policy and in supporting it. Macpherson (1979) reports that 61 per cent of the Oregon votes cast in 1978 supported the Oregon Land Conservation and Development Act 100, but that a follow-up sample of non-farm voters saw 80 per cent supportive of exclusive farmland zoning. This undoubtedly reflects the values placed on farmland by non-farm city people for amenity value as well as food production values, a phenomenon even more highly developed in Western Europe and especially in the United Kingdom (London Countryside Change Centre, 1989).

Some non-farmers have also played an important role in the formulation of farmland conservation policy through their involvement in *special interest groups*. Often loosely organised and with little money, special interest groups often achieve a substantial public profile. Good examples are the 1000 Friends of Oregon (1000 Friends of Oregon, 1980) and Ontario's Preservation of Agricultural Lands Society, more commonly known as PALS (Gayler, 1979). We shall develop the latter example. PALS was formed in the mid-1970s to oppose the extension of urban area boundaries in Ontario's Niagara Peninsula (Gayler, 1979; Jackson, 1982), and was later involved in the debate over the municipal plan for the City of Brampton. The objections of the group to the urbanisation of good quality farmland are essentially ideological. There is some evidence to suggest that PALS does not enjoy much support amongst the farming community. For instance, in his discussion of the group's attempts to preserve tender fruit soils in the Ontario Niagara Peninsula, Keating (1986, 35) comments that the group has 'been fighting a series of border skirmishes with farmers, developers and municipalities'. In fieldwork undertaken by Johnston (1989) near Toronto, many of the farmers contacted refused an interview until they were given assurances that the research was not supported by or in any way associated with PALS.

The fact that a group such as PALS has achieved such a profile raises a number of important questions. To what degree are they and their views representative of the broader public views – sometimes academic researchers have become involved and farm groups and developers have questioned this involvement of so-called third party 'observers'? How well-researched and scientifically sound are the positions they espouse (see, for example, Williams and Pohl, 1987)? Given that they have been the only non-governmental body to object to planning proposals in at least two major cases involving substantial erosion of agricultural land in Ontario, should we applaud them for their part in ensuring that these proposals were more thoroughly scrutinised?

The fourth group singled out here is comprised of *government agencies and departments*, including planners. This group has played an important role from time to time in providing information regarding farmland conversion, and individuals and occasionally groups have become involved with lobbying. In Canada the efforts of the Lands Directorate of Environment

Canada up until their recent demise constituted *the* only significant effort on a national scale (and for most regions for that matter) to establish the volume and quality of farmland losses to urban development (e.g. McCuaig and Manning, 1982; Warren and Rump, 1981). In the United States, the National Agricultural Lands Survey is also notable for its major contribution to a sounder information base regarding farmland conversion (e.g. NALS, 1981).

Planners play several roles. On the one hand, they provide important technical information to policy formulation. On the other hand, they can be very persuasive in their recommendations to the political decision-making bodies. They also therefore have a political role. In many respects, planners are just like everyone else; they have their own agenda, their own values, their own goals, their own limitations and their own prejudices (Munton, 1983a). It is unreasonable to expect planners to be absolutely objective in the execution of their duties. Here we are faced with a dilemma because the other actors in land use policy are forced into a position where their perspectives and goals are revealed. This is less often the case with planners.

Some planners defend decisions and recommendations on the ground of compliance with 'good planning principles' (Hodge, 1987). Similarly, development proposals can sometimes be refused by planning authorities because they are inconsistent with these same principles. The problem is that it is difficult to define what constitutes a set of 'good planning principles'. Deferring to 'good planning principles' to explain a decision simply hides the normative model and the value set upon which the decision is based. Because planners are in a position to exert considerable influence over land and land use in the City's countryside and elsewhere (Ratcliffe, 1974), it is essential that the theoretical and value frameworks they use in the execution of their duties be made explicit.

THE DEVELOPMENT OF FARMLAND PROTECTION POLICY: THE ONTARIO EXAMPLE. The consideration of the various actors involved in agricultural land issues emphasises the political nature of policy formulation and implementation. It is true both for the problem identification stage ('What is the issue? Who initiates or brings attention to the problem?'), the problem evaluation stage ('How important is the problem or the so-what stage?'), the development of an actual policy, its implementation and evaluation of the results.

While the state may play a key role in all stages and is theoretically accountable for any policies and strategies it develops, all the other actors can be involved to a varying degree at each stage. Recognising this means accepting that the search for any absolute objectivity is illusory. The priorities attached to the problems that will be dealt with by a government at any time are a function of several factors, including all the issues that face a government and its people and the relative power, influence and level of organisation of different actors and interest groups. The configuration of the policy will also be shaped therefore by a political process. We have thus to realise that what is a politically and culturally acceptable form of intervention in one jurisdiction may be a complete non-starter in another. Similarly, in evaluation (see below), political processes influence what interests are to be taken into account in answering questions such as 'Who pays and who benefits?' from particular policies and strategies.

The evolution of Ontario's farmland conservation Food Land Guidelines, adopted in 1978, illustrates several of these general points and provides a clear reminder that, for better or worse, the formulation and implementation of public policy is a very political process.

Prior to 1960 there was very little concern in Ontario about the conversion of agricultural land to non-farm uses. However, rapid urban expansion, combined with a sharp increase in the number of new lots being created in near-urban zones prompted the Ontario Department of Municipal Affairs to establish a set of criteria to assess development proposals and requests to create new lots in predominantly agricultural areas. These guidelines were not official government policy, they were simply designed for 'in house' use.

However, following the delivery of papers by Crerar (1963) and Hind-Smith and Gertler (1963) at the 1960 Resources for Tomorrow Conference in Montreal, an increasing number of observers began to question the wisdom of unchecked urban growth. Concern continued to mount in many rapidly growing areas to the point where the province's first official policy on non-farm development in agricultural areas was established in 1966 (Spooner, 1966). The policy, called Urban Development in Rural Areas (UDIRA), stemmed from the fact that many rural municipalities were allowing and even encouraging non-farm development only to realise that they were not generating sufficient new revenue from taxes levied on properties to support the extra cost of providing services to their new residents. Concern about the protection of farmland or any other special resource lands or unique areas were not part of the stated rationale of the UDIRA policy. An expression of anxiety over farmland losses was included, however, in a speech by the minister responsible for UDIRA in 1968 and again in 1975 (Johnston, 1983). Even though by 1975 a desire to protect farmland from urbanisation had become part of the rationale for controlling urban development in the City's Countryside, it was only one of a number of issues of concern, including minimum lot size (very much a local tax revenue issue), and the costs of servicing and the ease with which fire protection could be provided (efficiency of servicing issues).

Then, in the last quarter of 1975, a provincial election was held which saw the incumbent party returned to form a minority government. The balance of power was held by the only major party (the New Democratic Party or the NDP) who had raised the issue of the urbanisation of farmland during the campaign. In fact, in their campaign the NDP argued that between 1966 and 1971, 26 acres of Ontario farmland went out of production every hour (Manning, 1983). What the party either failed to realise, or failed to tell the electorate, was that most of the land in question was withdrawn from production because of changing economic conditions in regions far removed from large population centres (see, for example, Crewson and Reeds, 1982). Be that as it may, it is clear that the party holding the balance of power was the one favourably situated to influence government policy. It also bears repeating that it was the only one of three main parties in the election to express any concern over farmland losses and urban growth.

In March 1976, the government released a document entitled *A Strategy for Ontario Farmland*, in which it expressed its commitment to agriculture generally and to farmland conservation in particular. Two years later, after the release of a public discussion document, the *Food Land Guidelines* was tabled in the Ontario Legislature and formally adopted as guidelines for the provincial government (Johnston, 1983).

It is very difficult to assess precisely the role politics played in the formulation of Ontario's farmland conservation policy and the role of the various actors and groups in the process that led to the final statement. Researchers seldom enjoy access to backroom political discussions. However, why did it take the government of the day two years following the election to formulate their policy? Did the government's tardiness have anything to do with the fact that in their view the issue was not sufficiently serious to raise it as an election issue?

Public sector intervention in the market system

In this section, we briefly discuss the rationale for public sector involvement in agriculture, particularly in the City's Countryside. In a utopian system, it might be possible to envisage perfectly efficient resource allocation, in which there was convergence between private and collective or societal values and in which there was equity in resource allocation, rewards, and access to opportunity and public services. It might be difficult to define some of these concepts, but one thing most observers would agree on is that none of these qualities, however defined, is approached in reality.

Because of the difficulties that all market systems encounter, public sector involvement is widespread. Public sector involvement in agriculture generally as well as in the City's Countryside specifically arises because of the inability of the 'free market' system to address more collective values adequately. This public sector involvement occurs because of the presumed inefficiencies in the market system of allocation of resources, because of the divergence between private values and collective values, dramatically evidenced in many of the negative externalities associated with different types of development patterns, and because of the lack of equity associated with most development (who is really paying and who is really benefiting from different types of development?).

In addition to these interpretations of the need for public sector intervention, we also have to acknowledge the natural tendency of territorial political structures to adopt a protectionist stance towards their territory and constituents, partly because of solidarity with the constituents and partly because of a need to ensure their own continuity if the political structure is based upon some form of popular vote. This protective stance arises at all levels of government. At the local municipal level, municipalities compete for industrial and commercial tax assessment in order to maintain 'local' employment and tax base. They usually defend themselves vigorously against any developments apparently imposed externally which would transfer additional costs to their own populations.

The politicians representing provinces, states and regions become incensed when the farming sector in their jurisdiction suffers from high interest rates, poor international markets and the like, and endeavour to intervene or to lobby with central governments. Also central governments frequently behave protectively in their international dealings in order to defend and protect their own farm sectors; the European Common Market experiences during the 1970s and 1980s provide excellent examples of this, all within the framework of a move towards reducing international barriers to trade.

This general form of public sector intervention is directly related to intervention tailored to particular interest groups. In the situations discussed above, the interest groups are both geographically and sectorally defined. Concern arises because this form of intervention is not likely to be based upon a systematic weighing of all the values and interests potentially involved in any issue, and is therefore likely to be biased (Batley, 1983; Frankena and Scheffman, 1980; Jesson, 1987).

Finally, of course, on a much less constructive note from the perspective of broader societal interests is the public sector involvement that supports some form of self-interest on the part of the government structure in question, including the all too frequent conflicts of interest. We should also note the potential importance of the underlying ideologies held by key actors in the bureaucratic system who may try to impose or at least champion a particular cause or a particular method (Pahl, 1975, 1977).

The following discussion focuses on the rationale for public sector involvement regarding the inabilities of markets to deliver efficient and equitable systems of resource allocation. While much government attention has been aimed at these issues, frequently from the upper levels, we have to recognise that this involvement has usually received much criticism for being ineffective and inefficient as well! This is dealt with later when the different forms of government involvement are discussed and the role of local government involvement is highlighted in sustainable development approaches. Here, we first outline efficiency and equity considerations, and then we turn to the sources of market failure and provide examples for agriculture in the City's Countryside. Links are established with the three key dimensions of a sustainable food production system introduced in Chapter 1, viz. food sufficiency, producer rewards and stewardship of the land (Brklacich, 1989; Brklacich et al., 1991).

Efficiency and equity considerations

One by-product of the Industrial Revolution was the development of various social issues on a hitherto unprecendented scale and this marked the beginnings of massive public sector intervention in the market systems that accompanied the rise of industrial society (Zobler, 1962). Influential scholars like Marshall and Pigou argued that unconstrained markets led to inefficiencies and inequities which were detrimental to the general welfare of society (O'Riordan, 1971). Numerous economists have subsequently argued the case for public sector participation in resource allocation decisions (e.g., Gwartney, 1976; Lipsey, 1983; Rees, 1984).

Efficiency considerations are concerned with the relative costs of producing goods and services measured in terms of 'units' of output. Efficiency is a production perspective (Harvey, 1973). Under a perfectly competitive system, a free market will allocate resources in an efficient manner (Frankena and Scheffman, 1980). Resources will be allocated to different uses (goods and services) so that the returns (marginal returns) on the use of a given resource or input (e.g. land, labour or capital) between different uses are identical (Lipsey, 1983). Achieving this necessitates moving resources from one use to another as long as the returns on the extra (marginal) unit of resource is higher in the one use than in the other. The argument can be extended to many different resources or inputs.

An efficient resource allocation situation also implies that we are able to measure (or at least identify) all the costs involved in the production processes as well as measure the value of the goods and services produced. This immediately raises difficult questions when collective or public goods and services are identified which do not enter wholly or at all into the market system, in which value is expressed through market prices based on people's willingness to pay for a good or service (Feick, 1991; Musgrave, 1959; Samuelson, 1954, 1955).

Equity considerations cover the access to and distribution of rewards in society. Equity involves consumption (Harvey, 1973) which is often not in harmony with efficiency. It is not a simple concept to define, as it is related to notions such as 'social justice', 'fairness' and 'quality of life' (Merget, 1981). Equity has also been defined as equal opportunities and equality of 'life chances' (Rich, 1979). Equal opportunity also therefore implies equal accessibility to opportunities (Feick, 1991).

Equity concerns arise from many operations of the market system. One issue that has been at the core of many of the debates surrounding agriculture and the viability of agricultural production in the City's Countryside is the relative rewards earned from farming compared with non-farm sectors. This is the 'parity' issue which pervades practically all questions dealing with the long-term sustainability of agricultural production. Another pervasive issue is the distribution of the real costs and benefits associated with particular developments. For instance, how are the costs associated with exurban development shared? Are those who bear the cost most able to bear that cost? Who bears the real costs of the intensification of agricultural production in industrial farming? The same type of equity question – who pays versus who benefits – must also be asked of any public sector intervention in the market. Equity concerns arise partly because of the inadequate measurement of real costs (private and collective) in the market and partly because of the varying capacities of different individuals and groups and segments of society to ensure their needs are addressed effectively and fairly.

Inefficiencies and inequities arise because of the inability of the market to address all the values associated with the production of goods and services, both private and public. This inability is one of the driving-forces behind public sector intervention in the market system. The following discussion deals with three issues, externalities, other market imperfections, and equity concerns.

Externalities

Externalities represent the unpriced effects of supply; they have been variously referred to as 'third party' or 'spill-over' effects (Feick, 1991). It is important to emphasise that externalities can be either positive or negative, and can occur over space and across time. Externalities involve actions and events based on decisions – often at a specific location or set of locations – which generate side-effects or spill-over effects that affect other people's 'enjoyment' without them having been considered in the decision (Table 5.2). Externalities are therefore frequently associated with values that are only imperfectly, if at all, taken into account in the market system of exchange. Planning and management attempt to examine various externalities, but this intervention raises the very difficult question of measurement or how to introduce such values into management and planning.

NEGATIVE EXTERNALITIES. Negative externalities involve costs that are imposed through some action on someone who was not part of that action and who receives no compensation for the costs incurred. In the City' Countryside, examples of such negative externalities abound. Non-farm residents, for instance, who are irritated by noise from farm machinery or livestock odours coming from a nearby stock farm can be said to be subjected to negative externalities. They are not party to any economic transactions between the farmer and the purchaser of farm produce (except in a very marginal way), and hence do not enjoy any of the benefits. However, they are subject to some of the negative consequences of farming. Having a view from a residence 'spoiled' because of the construction of a new farm building or the removal of a woodlot would also qualify as negative externalities imparted to the exurban resident. These types of negative externalities complicate the issues of the sustainability of farming because they introduce non-agricultural values – and goals – into the debate (Table 5.2). Indeed, for some people like Crosson (1989), the negative externalities associated with farm technology and impact on water pollution and non-agricultural functions of agricultural land such as supporting open space amenity are more important than the concern over the loss of productive farmland to urban uses. The difficulty is that there is a lack of arrangements to compensate farmers and landowners for 'performing' the service of conserving these values.

Similarly, it has been argued that farmers can be the recipients of negative externalities from non-farm residential development, e.g. increased trespass and vandalism. This type of negative externality can be related to the

Table 5.2 Rationale for public sector intervention in agricultural land in the City's Countryside and the sustainability of agriculture

Considerations for public sector intervention in the market	Examples of related impacts in the City's Countryside		Sustainability of food production system: examples of linkages
	Felt by non-farmers	Felt by farmers	
Externalities			
Negative	Farm machinery noise Dust Livestock odours Aesthetics Water pollution		Potential impact on producer returns (negative due to complaints)
		Trespass Vandalism Severance and farmland fragmentation Traffic volume	Potential impact on producer returns (negative impact on operating efficiency)
	Road safety		
Positive		More solid community base due to exurban development Direct sales enhanced	Potential impact on producer returns
	Property upkeep	Property upkeep	
Intergenerational externalities			
Negative	Removal of farmland production capacity		Potential impact on long-term food sufficiency
	Soil degradation due to farm technology	Reduction in production capability	Potential impact on: food sufficiency stewardship producer returns
Positive		Adoption of soil conserving practices	Potential impact on: food sufficiency stewardship producer
	Conserving private		

	(landscape, property values)	
Other market imperfections	Quotas, subsidies, import controls, etc.	Positive impacts on producer returns Conflicting impacts on food prices Potential for restricting food supply
Equity concerns	Cost of public services (taxes, assessment, etc.)	Potential impacts on producer returns
	Increased awareness of farm/non-farm income differentials	Impact on stewardship (e.g. reduce costs), food sufficiency (e.g. if farmers abandon land) and producer returns (through opportunity costs)
	Access to countryside	

producer returns dimension of the sustainability of farming debate (Table 5.2). However, as noted in Chapter 3, while such conflicts between farmers and non-farmers in the City's Countryside have received considerable attention, particularly in the popular media, there has been a tendency to overestimate their seriousness both for farmers and non-farmers. It is interesting to note that some negative externalities can fall back upon the original decision-makers. A good example would be the farmer who sells off several parcels of land for non-farm residential development, and then subsequently finds him or herself faced with difficulties in using an awkwardly shaped field. Another example would be the creation of traffic safety hazards as the result of the development of several exurban residences in an area.

POSITIVE EXTERNALITIES. Most people would not complain about positive externalities! Positive externalities are benefits that are conveyed to a third party from an action in which the third party had no involvement. In the City's Countryside, examples of positive externalities for farmers as a result of exurban development include a more stable population (and market) upon which to base local service provision, compensating for example for the decline in the agriculturally supported population and population base for schools. Indirectly this contributes to the long-term sustainability of the farming system if it reduces servicing costs and creates a more (socially) viable community (Table 5.2). In terms of exurban residents, upkeep and improvement of both farm and non-farm properties confers benefits to neighbouring residents and property owners. In this case, the benefits may even enter into the market exchange system by influencing property values for adjacent properties.

Intergenerational externalities

The externalities discussed so far are felt by the third parties involved with very little time-lag from the time the action or decision is taken. However, other externalities accumulate over a much longer period, characteristically an intergenerational time-frame. Decisions taken today primarily to benefit the current generation (or, more realistically, part of it!) produce externalities for future generations. These externalities present a much more serious challenge to researchers and policy-makers.

NEGATIVE INTERGENERATIONAL EXTERNALITIES. These include decisions and actions taken today which will impose costs upon future generations and which have not been taken into account in arriving at the decision or action. There are important examples relating to agriculture in the City's Countryside.

The most widely debated example is a concern that underlies many of the farmland conservation debates. This has already been introduced in Chapter 2. The concern is that using land today which is well suited for agriculture will impose costs on future generations; it is inherently a resource scarcity issue. These costs may be food shortages, rising prices and a restriction of future opportunities because of the quality of the land. The concern is linked to food sufficiency and stewardship of the land in the sustainability of farming (Table 5.2). As already implied earlier, there is deep division of opinion on the issue, with some commentators expressing what can only be described as moral outrage over the inability of the market to take the interests of future generations into account (see, for example, Williams and Pohl, 1987), while others argue that there is little basis for concern because should food supplies start to become tight then the market will respond accordingly and make sufficient agricultural land available (Gardner, 1977). The conservationist approach has been dogged by surplus agricultural production in many Western countries, particularly in the European Community (Blunden and Curry, 1988). Measurement problems have proved almost intractable because of the uncertainties and intangibles involved, although efforts that have focused on the productive capacity of the farming system using scenario analysis, have clarified several issues (Brklacich, 1989; Smit and Cocklin, 1981).

Another issue, noted earlier, which has occupied much attention is the negative impact of agricultural technology used in industrial farming on the long-term productivity of the soil resource (Brklacich et al., 1989; Tousaw, 1991) and on non-agricultural values such as wildlife habitat as well as on pollution of water supplies generally (Blunden and Curry, 1988; Toch, 1988). Because of the association between agricultural intensification and the transformation of agriculture in the City's Countryside, it is reasonable to expect that these concerns would have been well articulated in the City's Countryside. In North America, this has not generally been the case, in contrast with much of Western Europe. The links between these long-term impacts and sustainability issues, both agricultural and non-agricultural, are clear.

POSITIVE INTERGENERATIONAL EXTERNALITIES. Conversely, positive intergenerational externalities are benefits, e.g. in the form of reduced costs, that future generations enjoy because of some present decisions or actions that have not explicitly taken them into account. This therefore excludes from consideration decisions taken with the intention of creating or enhancing such collective benefits. This eliminates public sector or interest group intervention to benefit future generations from this category. However, accepting the fact that conservation-oriented decisions or actions on the part of individuals also frequently involve an explicit concern for the future, we can none the less include here decisions by individual farmers to engage in soil conservation techniques. A non-agricultural example that falls more clearly within the definition is a private decision to maintain a particular landscape for personal enjoyment which has the result of conserving a 'culturally' valued landscape as well (Platt, 1972).

Other market imperfections

Other factors can also cause markets to operate in imperfect ways. For instance, imperfections can arise because

1. information on the relative merits of alternative courses of action is imperfect;
2. factors of production are not free to move in response to market signals (e.g. agricultural land is locationally fixed, union rules may restrict labour mobility, national boundaries may restrict cross-border movement of agricultural labour);
3. monopolistic or oligopolistic interests may restrict competition (e.g. vertical integration of food-processing interests into primary agricultural production through ownership or contract integration may reduce opportunities for entry into particular enterprises); and
4. government policy obscures market signals (e.g. price supports and supply magagement systems in agriculture).

It is somewhat ironic that some government intervention is undertaken to redress market imperfections which may have been the result of government intervention in the first place. This happens frequently when governments undertake planning and intervention exercises along sectoral lines which fail to take into account the complexities not only of the farming system but also of the whole social and economic system. The protracted debate in the European Community over the cost of agricultural support, overproduction and policies to cut costs and overproduction such as set-aside programmes is a classic example of this.

Equity concerns

We have already noted that equity concerns are quite different from efficiency considerations, but they are another factor of government intervention in the economy (Ervin et al., 1977). While equity concerns regarding income distribution (the distribution of rewards from the economic production system) and access to opportunity and public services has involved non-spatial policies aimed at redistribution through tax policy and social policy at the general level (Reynolds, 1966), there are specific concerns that are pertinent to agriculture in the City's Countryside which need to be noted, and which require different types of intervention (Table 5.2).

Equity of income differentials between farm and non-farm sectors is highlighted in the City's Countryside because of the stark realities of the differences that appear when farm and non-farm families live in close proximity. Such parity considerations underlie the reduction of agricultural labour much more so than does the demand for land for non-farm purposes. Dealing with this income issue is not an easy one, although as we shall see shortly, indirect attempts have been made to influence income through costs of production. Other approaches are essential, however, to address this issue.

Equity of accessibility to opportunities, e.g. through the educational system, and to public services, e.g. education, recreational facilities, libraries, social and medical facilities, has not been a major preoccupation of researchers in the City's Countryside. There are opposing tendencies, as has already been suggested. On the one hand, farming populations in the City's Countryside may benefit substantially where there has been an influx of exurban population because local populations may then be maintained or even increased beyond the population service thresholds required for particular services to be offered in the community (through the market and the private or public sector). On the other hand, exurban populations may increase the demands placed on the local municipalities and increase the costs to the community. While an expansion of the services offered can be seen as a positive feature, if the costs are distributed inequitably then there is further cause for concern (Orhon, 1982).

The challenge for public intervention

At this point, some important points are made regarding the main challenges that face government intervention in agriculture in the City's Countryside.

The discussion of public intervention in agriculture has underscored one of the key difficulties of such intervention: the collective interests that can be pursued present many conflicts among themselves as well as between different levels of government (local, regional, provincial/state, central state). In relation to this, five major points are made:

1. In order to analyse the situation in a given region, province/state and country, it is important to identify all the collective goals and interests that exist in the agricultural system and its component parts, to identify the conflicts between them and the levels at which such conflicts arise.
2. Sorting out the goals and collective interests being pursued – and they may not just be confined to agriculture – involves going beyond stated goals and objectives (Lowry, 1980). As implied in the earlier discussion, different actors or groups involved in the implementation of policy (e.g. local planning commissions, civil servants, etc.) may be influenced by their own ideas. However difficult, attempts need to be made to identify any hidden agendas (collective or private) that drive particular policies and strategies.
3. The collective concerns regarding equity (e.g. Who *really* pays for a given policy or strategy?) cries out for non-sectoral approaches and a concerted effort to trace the impact of policies and strategies ahead of time so that the number of unexpected impacts is as small as possible.
4. Especially in the mixed economies and democratic systems that characterise the developed world, those who create policies and programmes of government intervention must come to terms with the fact that the market system, with all its imperfections, still provides the major vehicle for the allocation of resources and the distribution of rewards in the functioning of the economy. Understanding how private interests (of individuals, firms) will respond to public intervention is likely to prove critical in designing effective programmes and policies. This is *not* because private interests are necessarily against the collective interest, but because they aim at ensuring the economic well-being of the individual and his or her family. Inevitably, failing to take the power of these individual motivating forces into account runs the risk of thwarting or at least rendering more difficult the implementation of policies and programmes to achieve collective goals and interests.

    The suggestion is that in many situations, public sector involvement needs to recognise that what it hopes to achieve collectively depends in part upon the co-operation of private interests. Government may try to regulate or it may be more persuasive (see below), but ultimate success depends in part upon private interests accepting such intervention. Often, the aim of public sector intervention may be to create a more favourable enabling environment in which private actions are influenced but in which the incentives for effective management and innovation, two of the essential ingredients for a viable and dynamic agricultural system, are not removed.
5. Finally, and in a similar vein, given the strong tendencies towards decentralisation of public sector responsibilities in many countries in the

last quarter of the twentieth century, a major challenge will be to ensure that collective interests and values that extend beyond the limits of local and regional jurisdictions and which can be affected by the negative externalities resulting from local and regional developments and decisions, are somehow incorporated explicitly into the local and regional level of planning and management. This means strong citizen and political support locally for strategies and policies (Conoway, 1990). This involves much more than public consultation and information dissemination, which are already widely-used components of planning in developed countries (e.g. Baschung, 1987), and necessitates more innovative approaches to integration of the various interests (see Chapter 7).

Our discussion has concentrated on planning and policy considerations, and this is an appropriate point at which to examine the different approaches and techniques used regarding agricultural land in the City's Countryside, paying particular attention to intervention aimed at farmland preservation and conservation.

Approaches and techniques of public intervention relating to agricultural land in the City's Countryside

The analysis of collective interests in agriculture, especially agricultural land suggests a threefold categorisation of interests and goals:

1. those related to the agricultural production of the land;
2. those which relate to non-agricultural functions of the land, while still retaining the agricultural use of the land; and
3. those which involve the land ultimately being transferred out of agricultural production to another function and use (Bryant, 1986b).

From the perspective of land planning, it is therefore necessary to distinguish between the *function* of the land and the land use *activity*. Function refers to the broad goals which land can be called upon to perform; activity refers to the actual use to which the land is put. The activity may well simply represent a means of managing the land to achieve its primary function. For instance, let us say that the primary function of an area of land is to contribute to flood management or supporting an attractive landscape. In both these cases, agriculture may still be the principal activity carried out on the land, but how it is practised is critical in determining whether the primary functions can be achieved. Agricultural structures built in a floodplain may impede water flow and hinder flood control. Similarly, ploughing an area may destroy an important element of a wildlife habitat, or modern farm structures (e.g. silos) may reduce the tourist and landscape value of an area. On the other hand, maintaining a particular landscape may require a viable agricultural system, and this in turn may require compromises between landscape values and the imperatives of a viable agricultural system.

With this important distinction in mind, how have the issues related to the agricultural land resource in the City's Countryside been approached? What sorts of techniques have been used to achieve whatever collective goals and values have been identified as important?

Approaches to public intervention relating to agricultural land

It should be evident that the agricultural land 'problem' is not a constant, either geographically or temporarily. The relative importance of the issues we have discussed varies from place to place partly as the whole range of problems requiring some sort of public intervention varies. Similarly, the broader economic and social context of agricultural issues changes over time. Futhermore, the broad approaches taken towards the agricultural issues is also partly a function of the general acceptability of different approaches to planning and public intervention generally. What works in one place and at one time may not work at all in another place, or even in the same place at another time!

Four main types of general approaches have been identified (Bryant et al., 1982; Bryant and Russwurm, 1982): negative regulatory, persuasive regulatory (both macro and micro scales), positive regulatory and integrated-comprehensive (Table 5.3). Some general points are made about each now; when we discuss the specific techniques, these broad approaches will again be mentioned where appropriate.

Table 5.3 Approaches and perspectives to agricultural land resource issues in the City's Countryside

Approaches to land use planning in general:
1. negative regulatory
2. persuasive regulatory
3. positive regulatory
4. integrated-comprehensive

Perspectives on agricultural land:
1. agricultural land as a residual
2. agricultural land as a special resource (a highly sectoral focus)
3. agricultural land as part of a functioning agricultural system
4. agricultural land as a support of amenity landscapes (and other functions)
5. agricultural production as one function among many (a non-sectoral focus)

The negative regulatory approach was the earliest attempt to regulate land use, and there are still strong components of this approach today in all countries (cf. Rickard, 1991b). Its prime purpose is to protect public welfare and safety as well as private property rights. Early attempts at land use zoning provide excellent examples of this approach. The locus of 'control' was relatively local and intervention was really quite minimal, particularly since the emphasis was so much on the protection of private property rights. Unfortunately, it tended not to give much consideration to the externalities associated with change except in so far as property interests were concerned. Interestingly enough, with the moves towards decentralisation of responsibilities in the last decade or so in many countries, local authorities are again coming 'into their own', and the sustainable development movement is likely to reinforce this even further (cf. Melbourne and Metropolitan Board of Works, 1977). Therefore, in order for this not to lead to simply a rebirth of the negative regulatory approach, some significant conditions must be present (see below).

With the realisation that the negative regulatory approaches were not adequate to deal with the externalities associated with rapid urbanisation and the transformation of the economy, other methods began to be developed. Initially, responses in North America were towards a more persuasive regulatory approach, particularly involving the development of intermunicipal co-operative management. Many of the negative externalities associated with growth and development were recognised – even if they were not measured and evaluated in any systematic fashion. One interpretation of the rise of such structures in many regions covering an urban core and a surrounding countryside is that they represented a compromise between 'doing' something and the realities of the municipal and local powers which were jealously guarded. Similar structures are to be found in parts of Western Europe, such as the management structures for the Paris region in the pre-1960s. This approach also includes programmes for farmland conservation in which participation by the farmers is essentially voluntary, such as the California Land Conservation Act (Williamson Act) of 1965.

It also became apparent that persuasive regulatory approaches were not enough to tackle many of the problems associated with growth and development generally in urban-centred regions, and so a more positive regulatory approach developed. Already in the United Kingdom, the Town and Country Planning Act of 1947 instituted a system involving much more centralised land use planning, in which the negative externalities of development were supposedly taken care of by regional and central state structures which were somehow more objective (Whitby et al., 1974). We find similar systems being developed in France from the early 1960s, where the elaboration of a regional planning structure initially provided the central state with a vehicle to become involved very directly in relatively local matters (although certainly not in relation to agricultural land issues) (Alduy, 1983; Bryant, 1986b). In North America, such central direction has been rarer and more sporadic, although some notable examples exist (e.g. Quebec's Agricultural Land Protection Law of 1978 and Oregon's agricultural land preservation programme). Any movement in this direction has usually been criticised for a variety of reasons (e.g. Frankena and Scheffman, 1980; Gardner, 1977; Gramm and Ekelund, 1975). Given experience, such criticism is not entirely unfounded. Often, the intervention was not necessarily any more 'objective' than that of more local forms of government, stories of inefficiencies abound, and there was a frequent tendency for the intervention to be particularly sectoral in orientation to the extent sometimes of almost subverting any more comprehensive local planning (Vachon, 1988). This occurred both in relatively centralised states, such as the United Kingdom, and in much more decentralised jurisdictions, such as California and Ontario.

Is the salvation some sort of integrated and more comprehensive approach? In some ways, some of the developments towards a more positive regulatory form of intervention, such as the growth of various regional government structures in various countries (e.g. in France throughout the country and in Ontario around the major metropolitan centres), have already provided a step in that direction. Even those structures, however, have had difficulties in coping with many issues. It has become apparent that many of these positive regulatory approaches have left out one of the key ingredients in the equation in the search for an appropriate development strategy – the people.

So the integrated and more comprehensive approach is still in its infancy. It is likely to have to combine elements of both the positive regulatory approach in some matters and the persuasive regulatory approach in others. Far from the integration of voluntary and co-operative strategies in planning being a relinquishing of collective responsibilities, it is probably just the opposite. Sustainable development needs input from different segments of the community on a systematic basis: While it is naïve to expect a large portion of the population to participate at any one time, there does seem to be more interest in participating in shaping one's own environment during the last years of the twentieth century. It needs to incorporate persuasive approaches on the macro scale (between municipalities and between local authorities and senior levels of government) and the micro scale (e.g. more management-oriented processes to dealing with land use issues) in a manner similar to that reported by Kerstens for rural development projects in the Netherlands (1989). Indeed, it could be argued that the success of any positive regulatory strategies, involving the development and implementation of rules and regulations, needs to be based as well upon a successful persuasive and co-operative approach!

It is interesting to reflect upon the linkages between these four approaches and the structure of responsibilities in a governmental sense. When the framework was first suggested, it was tempting to make a clear link between highly decentralised governmental structures (fragmented, uncoordinated?) and negative regulatory approaches to public intervention on the one hand, and more centralised structures (more objective, better co-ordinated, with the best interests of the public welfare at heart?) and positive regulatory approaches on the other hand. The implication was that a strong central (or regional) government would be in the best position to resolve issues and take a more objective approach to them. However, with mounting criticism of senior governments in all types of structures regarding the efficiency and effectiveness of their interventions, pleas for more decentralised structures have been heard with greater frequency. Indeed, if we are to subscribe to the evolving paradigm of sustainable development, greater responsibility for the local and community level is essential. Hence, the note in the previous section regarding the challenge of incorporating collective values explicitly into the local and regional planning and management.

Another element of a general framework of approaches to agricultural land is how agricultural land itself is viewed. Bryant and Russwurm (1982) identified five perspectives and it is useful to review these here (Table 5.3). The view of agricultural land as a *residual* after other uses have been allocated can be associated fairly easily with the negative regulatory approach, particularly since it meshes well with the idea of protection of private property values. However, even in more positive regulatory approaches with more centralised perspectives and greater attention paid to externalities, agricultural land is often still seen as a residual, because the areas in question are simply not seen as significant in relation to other planning problems. Bunce (1991) argues that in the highly decentralised planning systems of North America, agricultural land use 'designations' in metropolitan areas still tend to be seen as 'holding' categories for development purposes, a phenomenon related to the view of agricultural land as a real estate commodity rather than as a resource (Bunce, 1984).

At the opposite end of the spectrum is the perspective of agricultural land as a *special resource*. Agricultural land becomes an 'absolute', to which other issues and decisions must play second fiddle. A corollary at the local and regional level when agricultural land has been given special status is the institution of exclusive agricultural zones (American terminology, but the idea is widespread), in which the assumption is that any mixing of agricultural and other uses should be avoided.

Agricultural land as *part of a functioning agricultural system* incorporates a *business* perspective, but also the more general one of seeing agricultural land, if it is to be preserved or conserved, as part of an economically and socially viable agricultural system. This does not mean that only full-time farming is appropriate. Of course, some of these different perspectives are not mutually exclusive, but when one is combined with another, quite different results can arise. For instance, if we combine this perspective with the special resource perspective, we get the right-to-farm approach (Daniels et al., 1989; Lapping and FitzSimons, 1982; Lapping et al., 1983; Penfold, 1990). Farming practices take precedence over other residents and priorities in an agricultural reserve, outlawing complaints about 'normal' farming practices from non-farmers.

A difficulty with this approach is that it has often been associated with full-time farming. Once we admit the viability of a variety of farming systems, then concern about minimum farm sizes are seen for what it really is – subjective criteria based upon a particular image of what agriculture is and should be. If we admit alternative modes of production, then it may mean we ought logically to consider certain non-agricultural activities as legitimate and as supportive of the farming and the rural community generally because of the diversification of incomes it entails. While this has been recognised by some planning circles, they are often quite quick to reiterate that such permissive rules could only be allowed in very limited circumstances (see, e.g. London and SE Regional Planning Conference, 1987).

The idea of legitimate links between agriculture and other uses is even more explicitly brought out in the *landscape amenity* perspective, underlining the fact that collectively the major function for the land may be as a landscape support while the use is agricultural. This idea can be developed both in terms of the historic and cultural values that may be embedded in the agricultural landscape, the intrinsic aesthetic qualities of some agricultural landscapes, the supportive links between agricultural landscapes and natural environment (e.g. wildlife habitat) and the fact that agricultural land may also be an essential component of the overall landscape composition in an urban-centred region (see, for instance, the role of agriculture in the context of Green Belts).

Finally, in the *comprehensive* perspective, agriculture is seen as one function and activity among potentially many others. It implies a more integrated approach. It is not so much that the result may automatically differ from strategies based upon the other perspectives, but rather that reaching the policies and strategies selected involves the consideration of a broader set of costs and benefits associated with alternative futures for an area. We are probably closer to this in some of the regional level structures found in a number of countries than in any central state interventions.

Techniques of public intervention to achieve goals related to agricultural land

Before we concentrate on techniques relating to agricultural land in the City's Countryside, it is well to emphasise that the way in which agricultural land there can be shaped by public intervention is partly related to intervention in the agricultural area *per se* and partly to intervention in the urban area. Our focus is on the former, but some comments are needed on changes in the urban area. We must also distinguish between broad policies and strategies and the more specific techniques of intervention.

Because the City's Countryside is part of the broader functioning economic and social system of the regional city, strategies and techniques that influence the density and design of urban development also have ramifications for the countryside. On a broad scale, interventions must be seen in the context of any broader policies and strategies implemented to manipulate the urban area itself. Various methods have been used, including the delimitation of firm Urban Growth Boundaries (e.g. Pease, 1983). Other related approaches include Green Belts, Green Wedges, the Green Heart of Randstad Holland (van Oort, 1984b) and similar types of controlled or protected areas, as well as satellite urban centres or New Towns. Often, such broad strategies have been put in place to achieve particular objectives relating to Urban Growth Management.

Attempts at containment of urban growth through Green Belts and the accompanying attempts to densify development in the urban areas have been frequent, the most notable being in the United Kingdom especially in the London region (Elson, 1986; Hall, 1988; Hall et al., 1973; Mandelker, 1962; Munton, 1983a). Manipulating service provision to urban areas and transportation investments have also often been key elements (e.g. Furuseth,

1982) as well as managing the 'release' of land for servicing and lot production for urban uses (e.g. Department of Environment and Planning (Sydney), 1984). In all these approaches, being able to identify reasonable urban growth boundaries is important; this has not proved particularly easy to do, as the experience in the Niagara Fruit Belt of Ontario has shown (Gayler, 1982).

Satellite centres or New Towns have frequently been vital components of policies and strategies aimed at reshaping the urban area (e.g. Clawson and Hall, 1973). Such centres not only offer the potential to absorb 'overspill' population from the main metropolitan centre, but also to contribute to a more decentralised employment structure and a more efficient transportation network (Bryant, 1990). While many of these attempts have not yet met the objectives set for them initially, none the less they have had important implications for agriculture in the surrounding regions.

Specific techniques of intervention in the urban area regarding physical land use measures and development control (e.g. land use zoning, density criteria, coverage limits, lot sizes, height restrictions, servicing standards and open space requirements), manipulation of the rights-in-land (e.g. transfer of development rights, the use of expropriation, strategic public purchase of land and easement purchases or leases), and fiscal measures (e.g. property taxation, capital gains taxation) all have a potential indirect impact upon the countryside and its agricultural areas.

In relation to *intervention in agricultural areas*, we must again distinguish between the broad policies and the specific techniques of intervention. Where broader policies exist for the City's Countryside and for its agriculture in particular, they reflect the general conceptualisation that has developed for such areas (Bryant, 1986b). This can be linked to the dominant functions that agricultural areas are expected to perform (Table 5.1) and the perspectives taken regarding agricultural land (Table 5.3). Earlier in this chapter, these functions were divided into three categories. The following section analyses the agricultural production function identified in Table 5.4 and this is followed by a briefer consideration of the other two categories. In both cases, the framework presented in Table 5.4 is used to place the discussion of techniques of intervention under the headings of physical land use planning, rights-in-land, cost-influencing interventions, and information and management techniques.

THE AGRICULTURAL PRODUCTION FUNCTION AND FARMLAND CONSERVATION. At the outset, attention should be drawn to a rather troublesome terminological issue, farmland conservation or preservation. The two terms are frequently used interchangeably, particularly in the North American literature. 'Preservation' in the strict sense refers to absolute protection while 'conservation' involves protection without necessarily outlawing other types of development. Conservation is a more dynamic concept and we prefer to use this term because of this and because it can be incorporated readily into the integrated-comprehensive approach and the sustainable development notion.

Table 5.4 Techniques of public intervention in relation to the agricultural land base

Primary function	Physical land use controls	Rights-in-land	Cost-reducing	Information and management
Farm production	Land use zoning – local – regional – broad brush Severance controls Distance separation Right-to-farm	Purchase Expro'tion Transfer of development rights Purchase of development rights Lease or contracting-out of development rights	Financial assistance for mitigation measures Property tax rebates Use-value assessment Roll-back taxation Agricultural districts	Farm business counselling
Non-agricultural functions, e.g. Recreation Other resource production Other conserving functions	Land use zoning Regulations on structures and standards	Easements Land Trusts Access agreements	Financial assistance for coping with negative impacts	Information and management counselling
Future non-agricultural area (urban development, parks, etc.)	Land use zoning Timing Phasing of infra-structure Regulations on structures and investments	Purchase Exprop'ion Lease-back Relocation Retraining	Tax reductions if interim in farming Tax increases if wish to speed up development Relocation Retraining	Information on timing of development Counselling on relocation, etc.

To set the analysis of techniques into context, a brief review of some of the discussion from Chapter 2 concerning the evolution of the nature of the 'problem' is given next.

In Britain, concern over urban sprawl and its impact on the agricultural land base can be traced back nearly fifty years (Blacksell and Gilg, 1981). From the end of World War II until 1959, growth of urban areas in Britain was responsible for the conversion of an average of 38,200 acres of farmland per year (Best and Coppock, 1962). While these losses were actually lower in the post-war years than they were in the period immediately preceding the outbreak of hostilities, the rate of conversion was sufficient to spark concern. In the United States, farmland conversion first attracted attention in the mid-1950s (Bogue, 1956), while in Canada this did not occur until the early 1960s (Crerar, 1963; Hind-Smith and Gertler, 1963). In New Zealand, concern within the academic community can also be traced to the early 1960s (Stonyer, 1973), although legislative attempts to control the spread of urban areas into good farmland occurred as early as the 1950s with the passage of the Town and Country Planning Act of 1953. In 1956 the New Zealand Planning Board also held that land with high actual or potential value for food production ought not to be used for urban development in their ruling on the case *Blakely* v. *Manukau County Council* (Robinson, 1968).

In these places, as elsewhere, once the 'problem' of farmland urbanisation reached the public agenda, policy formulation was soon activated. Like 'the environment' in the late 1980s, political parties especially in North America both to the right and the left of the political spectrum expressed their disdain of uncontrolled urban growth and promised action. Policy to conserve farmland generally and in the City's Countryside in particular have today been adopted in many countries (see, for example, Trzyna, 1984). However, the United States and Canada are undoubtedly world leaders in terms of the number and variety of different farmland conservation strategies devised. Daniels and Reed (1988) estimate, for example, that in the United States over 400 counties have zoning regulations designed to afford a measure of protection to farmland. As of the early 1980s, farmland protection mechanisms have been deployed in North America by all ten Canadian provinces and all but two US states (Furuseth and Pierce, 1982b).

The details of the various approaches used have already been widely discussed in the literature, both in general reviews (see, for example, Cloke, 1989a; Coughlin et al., 1977; Corbett, 1990; Dawson, 1984; Furuseth and Pierce, 1982a, 1982b; Goldman and Strong, 1982; Held and Visser, 1984; Rose, 1984; Trzyna, 1984), and with respect to particular mechanisms (see, for example, Buckland, 1987; Furuseth, 1980, 1981; Hansen and Schwartz, 1975; Krueger and Maguire, 1984; Luzar, 1988; Moran, 1980; Phipps, 1983). Our analysis therefore centres upon the synthesis suggested by our framework (Table 5.4).

Physical land use measures and development control. Generally, the most important class of techniques which has been used to influence farming and to attempt to maintain agricultural activity consists of physical land use controls over the *location* of different land uses. The most common approach is to identify agricultural land use zones in official plans and zoning by-laws.

Agricultural land use zones have been developed on various geographic scales, depending upon the particular national or provincial and state context and are the most frequent response of government to farmland conservation. At the local level, the geographic prescription of land uses characteristically has involved the separation of agricultural uses and urban or non-farm related uses. Certainly in the early days of land use planning, this can be considered the example *par excellence* of the negative regulatory approach. Zoning of one kind or another is found throughout the developed world (see, e.g., Aitchison, 1989; Berteloot, 1972; Cloke, 1989a; Geay, 1974; Listokin, 1974; Nelson, 1977). The theory is that the distribution of land uses through zoning represents the articulation and partial implementation of broader planning at the community level, although this still remains elusive in many areas (Daniels et al., 1989).

The agricultural zone with no other activities in it is rare. Generally, while zoning may involve designation for a specific use, it is very common for a range of uses to be listed as permitted and others as prohibited. Furthermore, there are often regulations associated with a given zone category specifying the conditions under which a particular use can develop. It is not unusual for different types of agricultural zones to be designated, reflecting a combination of agricultural land quality and the existing land use pattern in an area (for an Oregon example, see Pease (1990)). It is also possible to talk of 'buffer' zones around critical core agricultural areas (the 'exclusive' agricultural zones) or particularly sensitive agricultural areas. These buffer zones depend upon the assumption, however, that there is a simple and very direct distance decay function associated with land use incompatibilities. Rather than buffer zones of less sensitive agriculture, effective barriers are sometimes created by real barriers to interaction, e.g. high speed and limited access highways.

Differences would typically relate to how tight the control would be on non-farm development in such a zone. One of the most important failings of zoning has been its limited relationship to longer-term planning, especially in rural areas in the City's Countryside. It is under such circumstances as these that the negative regulatory approach is most often encountered, and it has been a simple matter for people to argue that protection of private property rights is synonymous with allowing land to be redesignated to 'higher and better' uses when the time is right. This shows how important it is to have effective planning policy and not just one that is preoccupied with land use *per se*. Where there is a long-term planning policy in place, then decisions on how much land is required for development and how much needs to be serviced are likely to be more realistic. Under such circumstances, defining the appropriate limits for at least any accretionary development is as important as identifying the zones for predominantly agricultural use.

Agricultural zoning of a somewhat different kind also exists in some places on a wider scale, e.g. that of the province or state in North America. These are not the broad zones that one might expect in an official land use plan oriented towards the long term. These are broad zones or reserves for agriculture which have been developed with a significant amount of centralised direction (e.g. Hawaii, British Columbia, Quebec). The assumption for this broad brush approach is that it is impossible to evaluate adequately at the local level matters which are of regional and national importance. They provide therefore very good examples of the positive regulatory approach. Of course, they have usually involved the local and regional levels of government (where the latter exist) in integrating these broader zones into the more local processes, in receiving input or even in actually developing the reserves or zones under the watchful eye of the upper level of government.

Where physical integration of the broad brush and local or regional level plans does not occur, then the broader zones take precedence over the local plans and by-laws, e.g. as with the Quebec Agricultural Land Protection Law

(Debailleul, 1988). Elsewhere, even where there is no explicit agricultural land protection 'law' but where there is a strong relationship between national and local physical planning as in the Netherlands (Grossman and Brussaard, 1987, 1988), land use zones for agriculture at the local level can reflect strong national commitments to continued agriculture. Once again, however, in the long term it appears that to be effective account has to be taken of urban growth and economic development patterns to establish realistic urban growth and development boundaries. Furthermore, it is also essential to have sound information available when delineating such reserves.

These more centralised agricultural land use programmes have become a quite common farmland conservation mechanism, especially in North America. However, they have been implemented much more widely in Canada than in the United States (Furuseth and Pierce, 1982b). There are undoubtedly differences between Canada and the United States in terms of institutional and constitutional frameworks for planning, but also attitudes differ in the two countries toward government participation in the economy generally and land use planning in particular. Even so, there are significant differences within Canada: contrast the central Canadian provinces, Quebec and Ontario. Ontario has had a very persuasive regulatory approach on the macro scale, trying to persuade municipalities to implement suggestions provided by the provincial government to set aside prime quality agricultural land and so forth; while Quebec has developed a much more draconian system of control modelled on the early stages of the British Columbia Agricultural Land Commission (Debailleul, 1988; Vachon, 1988).

Where and when they have been implemented, the more centralised schemes seem to have been fairly successful in preventing the urbanisation of farmland (Furuseth, 1981; Furuseth and Pierce, 1982b; Thibodeau, 1984; Thibodeau et al., 1986), although the difficulties of objective evaluation are substantial particularly in taking into account the impact of changes in external conditions such as the state of the economy. In the case of the British Columbia Agricultural Land Reserve (ALR) programme, for example, it was shown that it can also have a positive affect on the general atmosphere for farming in a region (Manning, 1983; Manning and Eddy, 1978). However, as Pierce's (1981a) evaluation of the British Columbia scheme demonstrates, farmland losses can still occur. The success or failure of such broad brush zoning schemes rests heavily on the degree to which the body making decisions on applications for the development of non-conforming uses are committed to the preservation – or should it be conservation? – of agricultural land. In cases where the bodies ruling on the applications are comprised of politicians or their appointees, the success or failure of these sorts of schemes depend upon political leadership.

However, it is a fine point for debate about when the use of centrally conceived land use policies becomes truly positive regulatory in nature and when they are really a persuasive regulatory mechanism. In the Ontario case, even though the provincial government is ultimately responsible for planning, in a number of instances important responsibilities have been delegated to regional governments and even local governments. Ontario's Food Land Guidelines, referred to earlier, are most certainly classic examples of persuasive approaches on the macro scale and it would be very difficult at the present time to suggest that Ontario has anything approaching a broad brush policy towards the designation of agricultural zones. Quebec's agricultural land use policy is implemented partly through the local and regional municipalities, but the province has maintained until now considerable powers over the exclusion of land from the reserves – it is more certainly positive regulatory in orientation.

As a final example, New Zealand's Town and Country Planning Act states that in the national interest, planning schemes will, amongst other considerations, provide for the avoidance of the encroachment of urban development on, and the protection of, land having a high actual or potential value for the production of food, and the avoidance of unnecessary expansion of urban areas into rural areas in or adjoining cities. All planning schemes receive ministerial approval to ensure compliance with the act, and all development proposal and subdivision applications are assessed in relation to the schemes. This appears to be a much more positive regulatory approach than that of Ontario.

Another type of broad zonation of agricultural land, agricultural districting, is dealt with in the following section because it rests essentially upon voluntary participation and so fits rather uneasily with the other techniques being discussed which involve restrictions on the rights of landowners.

A final point should be made about zoning, particularly the local and regional kind. There are often other restrictions that go hand in hand with zoning, such as severance controls (e.g., Pease, 1982), density controls (e.g. Goldman and Strong, 1982), set-backs and right-to-farm laws. As one example, we comment on severance control or the control of the process by which properties are split into two or more legally independent parcels of land. They are known confusingly as subdivisions in Western Canada and parts of the United States, and lots in the United Kingdom. Severances (excluding multiple property division for residential, commercial and industrial subdivisions or estates) are created for several purposes. These include simple modification of property boundaries for convenience, road widening and building lots. Control over the severance process was assumed to provide a means of influencing the rate and nature of residential development in rural areas, particularly in North America where large property parcels were commonly the rule.

We must hasten to add that the introduction of severance control in most areas in the early stages had very little to do with conservation of agricultural land, and everything to do with local property taxation and the costs of servicing the more scattered forms of development. Restrictions ranged from setting minimum lot sizes to specifying the maximum number of severances that could be requested from an initial area of land. Minimum lot sizes were often set large, having the effect of potentially removing more land from agriculture than if small sizes were permitted, but probably more likely to ensure more wealthy owners – and the corollary, more expensive homes more capable of carrying their weight when it came to paying taxes! While severance control is talked of nowadays in relation to 'preserving' farmland or controlling the development of incompatible land uses in an agricultural area, local taxation and costs still loom large in many people's minds. For related reasons, in some situations the large minimum size of lot is also aimed more at maintaining attractive residential areas than at controlling development in agricultural areas (Goldman and Strong, 1982). Depending upon the jurisdiction, local municipality, regional municipality or county have been the key authorities in approving severances, frequently within guidelines established by upper levels of government.

Initially, severance criteria were not well co-ordinated, if at all, with zoning controls, making life quite tiresome for everyone concerned. Increasingly, there is co-ordination. However, given the relatively decentralised approval mechanisms usually involved, difficulties remain because the people on the local and regional approval bodies rub shoulders daily with the applicants (Meister, 1981) and sympathise with farmers' financial troubles (Gayler, 1991). For instance, difficult issues surface continually such as how to handle the severance of a lot for the farmer's son or daughter, or hired help, or for the farmer's own retirement. If the underlying philosophy of planning is to 'preserve' farmland as much as possible and 'keep out' non-farmers, then in the long term many of these severed lots would likely fall into the hands of non-farmers because of continuing farm consolidation in many areas. And that assumes the request for a severance has been stated honestly! Allowing construction of the houses concerned without a severance, thereby tying the sale of the newer house to that of the whole farm, only seems to put off the issue. In some ways, the solution adopted in some municipalities of allowing the son, daughter, hired help or retiring farmer to live on the farm but in a mobile home, seems destined to create a sense of a second-class citizen. Attempts to link minimum sizes of lots to 'viable' agricultural units raise as many questions as they answer, particularly the underlying image of what constitutes 'good' farming (Pease, 1982).

A closely related example is the introduction into the regulations for agricultural zones – where some development is almost always bound to rear its head – of requiring minimum distances between intensive animal husbandry farms (e.g. feed lots, broiler chickens, pig production) and any new residential development – and vice versa. In some areas, because of the density of animal operations this is almost tantamount to zoning for exclusive agricultural use on a large scale! In Ontario, where some municipalities have introduced such regulations based upon the Agricultural Code of Ethics (Ontario Ministry of Agriculture and Food, 1976), the application of the principle seems to be highly scientific and technical. According to the nature of the farm and its size, tables have been provided showing the minimum distance at which any new residential development should be allowed. However, underlying this apparent technical solution is a set of very subjective assumptions.

The types of techniques discussed in this section all include restrictions on the rights of owners over their properties. Even where some of those techniques, such as set-backs and minimum distances, are used for a development control or performance zoning (Cloke, 1989b; Blacksell and Gilg, 1981; Dawson, 1984; Pease and Morgan, 1979; Stockham and Pease,

1974) – you can develop this land use, but only subject to these conditions -- there is still a direct control being exercised over landowner's rights in land. *Rights-in-land.* In this section we have included techniques that deal with rights-in-land where there is no taking of a right – or if there is, there is some direct compensation offered.

Techniques that respect owners' rights over land involve the purchase or lease or other form of transfer of an owner's bundle of rights or a specific right in land. Only in the case of expropriation is there any coercion involved; and while the landowner experiencing expropriation may not be pleased with the situation, generally he or she is well compensated. In all these techniques, it is possible to say that the intervention still accepts the market as a legitimate mechanism for determining the value of land. These techniques can therefore generally be considered to be persuasive regulatory (only the expropriation case is questionable here).

Outright purchase of the title to land on any large scale is rare. An exception in North America is the Ottawa Green Belt, where over 20,000 ha of publicly owned land constitutes an enormous asset in the region; of this, over 14,000 ha were acquired by the federal government expressly for the Green Belt and is now owned and managed by the National Capital Commission (National Capital Commission, 1982). However, while there is much agricultural land within the Green Belt, conserving agricultural land was not the prime purpose of the purchases. Rather, it was to be used to contain the urban growth of the capital city and contribute to the quality of life there. It is interesting in this respect to note that today there is considerably less agriculturally used land in the Green Belt than when it was started after 1956. Obviously, public ownership of farmland in no way guarantees continued agricultural use of the land!

Expropriation of agricultural land for agricultural purposes and use is extremely rare. Examples include agricultural land that was purchased in the course of land assembly for New Town development in France where the continuation of agricultural activity contributed to the landscape composition of the town – but this takes us to non-agricultural functions of land.

Strategic purchase of agricultural land destined to remain in agriculture as a means of conserving larger tracts of land is, of course, always a possibility. A variant of this, not frequently used in the urban fringe, is the pre-emptive right to purchase farmland that the French SAFER (Société d'Aménagement Foncier et d'Etablissement Rural) are able to exercise to buy up land in an 'agricultural' zone and use to help restructure other farms in the zone (Bryant, 1975).

This brings us to the question of the purchase or transfer of development rights, one subset of the bundle of rights that reside in ownership of land. The transfer of development rights may involve the farmland owners selling, leasing or otherwise agreeing not to exercise the right to develop their land in exchange for some form of compensation, typically cash (Buckland, 1987); in some cases, it actually promotes a free market in development rights (Manning, 1983). Adopted in five US states at the time Furuseth and Pierce (1982b) completed their study in the early 1980s, this farmland conservation mechanism has yet to be adopted anywhere in Canada. Rawson (1977) argued that the reluctance of Canadian planners to embrace the idea is because in Canada landowners are really no more than tenants of the Crown, with landownership resting with society in general and not with individuals. The notion that separable rights in land can be transferred to the 'state' or some other agency and compensation made therefore undermines the basic concept of land tenure in Canada.

The major problem with the transfer of development rights to protect farmland from being urbanised is that it does not work very well (Nelson, 1990). For example, Pitt et al. (1988) found that farmers in areas subject to the most intense development pressures were the most reluctant to participate in the development rights purchase programme of the Maryland Agricultural Land Preservation Foundation. Given the use value of agricultural land which is likely to be developed, this finding is hardly surprising. Another problem with this strategy is that to be successful it is expensive (Esseks, 1978; Mooney, 1990). The theory of separating development rights is fine; but when the potential development value accounts for the bulk of the 'market value' of a property, and when legal costs are added, it is not clear that much is saved compared to outright purchase of the title to the property. In considering the possibility of establishing markets in development rights, the major shortcoming is the ability of planners to define the 'preserves' and the 'development' areas realistically.

A related technique is that of the densification of development rights in an area or, more commonly on a property. In France, for instance, the *coefficient d'occupation du sol* (COS) in a land use plan summarises the ratio between net built floor space and property size. A COS of 2.0 for example means that a building with a minimum floor space of 200 square metres must be built on a minimum property parcel size of 1,000 square metres. When the COS is set low, permitted densities of development are low. Its effect is not unlike some severance control procedures, except that the focus is on building, not property division.

When these COS have been laid out in a plan for some time, it is argued that they almost resemble an acquired right. Yet, in many instances, conditions have changed since the time many land use plans were initially formulated, and a municipality may feel that permitting development at the density 'permitted' would not serve its interests. It might for example give rise to low density development in an area where it is felt that open space, say for wildlife habitat protection, should prevail. Densification is a persuasive regulatory technique. It recognises the 'acquired' rights but drops the densities to zero on most of the property and increases it substantially on the remainder. Undertaken on several properties, it can lead to significant clustering of development and becomes an explicit transfer of development rights. To work, however, people have to accept and have faith in the various land zones designated – something of a tall order, given the collective experience.

Another example of this kind of modification is the potential development of residential-agricultural condominiums. It is similar to densification, except that the property portion with a COS of zero remains part of the property that is developed as a condominium. One could argue that in those countries where private property interests are jealously guarded, such an arrangement allows farmers to sell, reinforces clustered development and stands a greater chance of maintaining the land as open space – and in agricultural use if the collective owners can rent it out to farmers – than if it were in public hands, e.g. a municipality. This is not unlike the agri-community examples discussed by Mooney (1990).

A final example of the separation of development rights, albeit usually temporarily, which is also persuasive regulatory and is a cousin of the broad brush zoning referred to earlier, is the creation of voluntary agricultural districts (or preserves). It is essentially a strategy that has been developed in the United States (Furuseth and Pierce, 1982b), first in New York State (Conklin, 1980). Membership in agricultural districts is voluntary and minimum sizes on farms that can apply to be placed in an agricultural district are typically set. Incentives are often used to encourage farmers to form a district, such as tax breaks which are linked to district membership.

Where they have been formed, agricultural districts do seem to be associated with stable agricultural land. However, this mechanism suffers from the same basic problem as the purchase of development rights (from the perspective of those advocating conservation of farmland). Farmers in a position to sell their land for urban development seldom fall over themselves in the rush to form an agricultural district! The reasons are not difficult to understand, for despite the utterances of many farmers about the need to retain good quality farmland (Molnar, 1985), the equity in their land represents the retirement nestegg for many of them. The incentives are not that great. So who can blame them for wanting to retire in comfort without having to worry about bills after years of back-breaking farmwork? However, Napton (1990) claims greater effectiveness for the Minnesota Metropolitan Agricultural Preserves Act of 1980 because there is a much closer link between the state and local planning, reflecting an integration between land use zoning (long-term agricultural land) and eligibility to participate in the programme. Farmland owners who agree to participate sign a restrictive agreement to retain land in agriculture for eight years and to use sound soil conservation practices. The automatic renewal provision is similar to the California Agricultural Conservation Act.

Cost-influencing interventions. Another major set of techniques used to retain farmland aim at lowering the costs to farmers of staying in agriculture (e.g. property tax reductions) or increasing the costs on land which is needed rapidly for development (e.g. increasing property taxes). These again operate within the rules so to speak of the market (Gardner, 1977). There are also examples of attempts to help farmers cope with some of the costs of operating in an urban fringe, such as subsidies to help install and repair fences, or setting up one-way road systems just for agricultural traffic in urban fringe municipalities to try to reduce the costs associated with farm machinery movement in fairly urbanised zones. However, the most important techniques to date in this general category of cost-reducing intervention deal with property taxation, and we shall concentrate on the annual property taxation from local and regional municipalities. The property taxation problem for farmland in the City's Countryside has received much attention in North America, and was related to the tendency for farmland to be overvalued compared with non-farmland in the urban fringe, and for rural–urban fringe municipalities to have higher costs for servicing because of the non-farm population (Gloudemans, 1971; Hady and Sibold, 1974; Hady and Stinson, 1967; Roberts and Brown, 1980). Farmland was therefore exposed to higher taxes in the rural–urban fringe than farmland elsewhere.

Regional standardisation of assessment procedures have eliminated most of the problems associated with 'unfair' valuation. So attention has been oriented to reducing the actual burden of taxes, and programmes using property tax incentives to slow the rate of farmland conversion have become very commonplace in the United States (Forkenbrock and Fisher, 1980; Nelson, 1990; Skoretz, 1990).

In some schemes land meeting certain criteria (e.g. the 'bona fide' farmer or farmland) is assessed on its use value for agriculture, instead of on its economically highest and best use, an approach known as 'preferential assessment' or 'pure' differential assessment (Nelson, 1990). A variation on this is deferred taxation or 'roll-back' taxation. By this, preferential assessment of farmland is the rule, but the forgone taxes over a specified length of time on the difference between the agricultural use value of the land and the assessed value it would otherwise have received becomes due, with interest, if the property is then sold or converted to non-farm development. Another less frequent variation is preferential assessment with a penalty assessed upon conveyance or change of use (Nelson, 1990).

In other cases, such as in Ontario, property tax on farmland meeting certain criteria is now waived altogether, but with provision for recovery if subsequently developed or reassessed for residential, commercial or industrial use. Furthermore, as Coughlin and Keene (1981) report, in the United States the use of property tax incentives has been extended to death duties. One of the provisions of the 1976 Tax Reform Act passed by the US Congress allowed estates taxes on land in family farms to be calculated on the basis of use value thus relieving the heirs of a deceased farm of the need to sell the farm in order to meet estate duties. In some Canadian Provinces and US States, provision exists for local authorities to exempt farmland from special levies to cover urban infrastructural expenditures such as the installation of trunk sewer lines and road upgrading necessitated mainly as a result of the growth of non-farm populations (Furuseth and Pierce, 1982b). Refundable income tax credits have also been allowed against property taxes exceeding certain income percentages for low income farmers in a few US states (Nelson, 1990; Skoretz, 1990).

In some cases, property taxation programmes have been literally at the core of a farmland conservation programme, as in the California Land Conservation Act of 1965 (Gustafson, 1977; Gustafson and Wallace, 1975; Snyder, 1966; Williams, 1969). Here, yet another persuasive regulatory approach involved farmland owners entering into a contract with local governments. This could occur in an already designated agricultural area or upon application by the owner. The contract effectively removes the development rights on such a parcel of land for ten years, automatically renewed annually unless notice to the contrary is received. In return, the farmland owner received preferential assessment for property taxes. Most of the land under contract was not in the areas of most intense urban pressures, as is to be expected under a voluntary programme (Carman, 1977). The New York Agricultural Districts in a similar vein involved roll-back taxation and preferential assessment (Bryant and Conklin, 1975), and again most of the take-up of the programme in the early stages was not in urban fringe areas.

Politically attractive, the elimination, reduction or postponement of taxes on property as a means of protecting farmland from urbanisation has worked, however, with only very limited success. Furuseth and Pierce (1982b) cite a number of *ex post facto* evaluations of property tax incentive schemes (see, for example, Barron and Thompson, 1973; Furuseth, 1980; Gustafson and Wallace, 1975; Schwarz et al., 1976; Vogelar, 1978), which conclude that this approach is not particularly effective in reducing the amount of farmland converted to urban use. There is evidence, however, which suggests that reducing or postponing property taxes on near-urban farmland may provide an incentive for land to remain in production longer (see, for example, Johnston, 1989), and therefore these policies can have a positive effect in lowering the amount of farmland set aside in anticipation of conversion. However, this is a different objective from that of reducing farmland conversion. In any case, the evidence is ambiguous (Nelson, 1990).

Information and management approaches. The final category of intervention deals with approaches which are also persuasive regulatory and which are aimed at helping farmers cope with particular urban fringe problems by providing them with advice on how to adapt to and take advantage of their situation. Experience is very limited here. To cope with urban fringe pressures, the urban fringe management projects of the Countryside Commission in the United Kingdom (Countryside Commission, 1976) offered some interesting methods with small sums of money being made available to farmers to help deal with some of the realities of their location, e.g. people using farmland for hiking. While management advice is provided in some areas where an agricultural extension service exists, frequently the agricultural extension workers themselves have not been especially trained in the types of enterprises oriented towards the urban market, so it is often sporadic and not part of a coherent programme. Ilbery (1988b) in his study of farm diversification in the Birmingham, United Kingdom, urban fringe in 1986 notes that relatively few farms had benefited from advice before diversifying, despite its availability from various agencies. He suggested a more co-ordinated approach to the issue. Underlining this plea, Rickard (1991a) suggests that the potential for direct marketing in Connecticut has yet to be realised because farmers on average have not yet grasped its significance. Therefore, the potential appears to be significant given the already reported trends in the market for agricultural produce. Mention should also be made of information (marketing) programmes aimed at the urban population, not just to sell agricultural production but also to sell the image of an agriculture that contributes to the quality of life of the whole metropolitan region (e.g. the activities of the Regional Chamber of Agriculture in the Ile-de-France, Paris).

DIFFICULTIES IN EVALUATING FARMLAND CONSERVATION POLICY. Before moving on to discuss briefly the other categories of goals and techniques of public intervention regarding agricultural land, some summary comments are offered on the difficulties of evaluating the success of farmland conservation policy and programmes in preventing the urbanisation of good quality farmland.

As in any policy and programme, evaluations can be undertaken at different stages in policy formulation and development (Anderson, 1979; Smit and Johnston, 1983). First, public intervention implies that there is a problem with which existing policy is unable to cope (Sewell and Coppock, 1976) and it is also usually a problem which has received significant public recognition (Frederic, 1991; Sewell, 1983; Wiseman, 1978). What is the real importance of the problem and has its significance been adequately gauged? Is it sufficient to warrant intervention? In earlier discussion we have already suggested that frequently the real nature of the problem in the area of farmland conservation has not been identified and certainly not quantified, and the issue continues to be clouded by surplus agricultural production and excess capacity in relation to the effective market in many Western countries (Crosson, 1989).

In the second stage, one or more objectives (preferably measurable) are then established, the attainment of which represents a solution to the problem. Is the link between the problem and the objectives well documented and established? Or will achieving the objective(s) really make a difference to the problem? If there is more than one objective, are they consistent with each other? How are they affected by other objectives of public policy in other domains? In general, measurable objectives stated as a target are non-existent for farmland conservation programmes; it therefore becomes very difficult to know when success is achieved! Almost inevitably, some farmland loss will occur whatever the nature of the programme – so the amount that is 'acceptable' is an important issue that cannot be ignored (Pierce, 1981a)

Johnston and Smit (1985) investigated the links between the first two stages of the policy formulation process by analysing the rationale for the Food Land Guidelines, the farmland conservation policy adopted by the Province of Ontario in 1978. Was the proscription of urban activities from agricultural areas (the specific objective of the Guidelines) both sufficient and necessary to ensure the long-term viability of Ontario's agricultural industry (the broad policy problem at which the Guidelines are aimed)?

An examination of the available literature identified six dimensions of the policy problem: one direct impact of non-farm development on agriculture and five indirect impacts (Table 5.5). Johnston and Smit concluded that the objective of Ontario's Foodland Guidelines was insufficient in itself to mitigate any of the dimensions of the policy problem, and necessary to mitigate only two of them, and then only in particular circumstances (Table 5.5). This is not surprising given the complexities of the factors affecting agricultural viability. The policy appeared to be an inappropriate response to the conditions that stimulated its formulation.

Table 5.5 Evaluation of the rationale for Ontario's Food Land Guidelines

Dimension of the policy problem	Objective necessary?	Objective sufficient
Decline of the agricultural land base	Yes (in some regions)	No
Increased farmland values	No	No
Increased farmland rental	No	No
Conflicts between land-uses and life-styles	No	No
Changes in rural economic infrastructure	Yes (in some regions)	No
Uncertainty for agricultural decision-making	No	No

Source: Compiled from Johnston and Smit (1983, 236)

In the third stage of policy formulation, one or more strategies will be developed to achieve the objective(s) selected at the second stage. This may involve the identification of alternative implementation strategies, an evaluation of each in relation to their potential contribution to achieving the policy objectives, and the selection of a preferred implementation strategy. This last consideration will often take into account such factors as budget availability, public opinion and political acceptability. Evaluation questions will deal with all of these to try to understand the selection. This is generally known as ex-ante or formative evaluation of a policy (Seni, 1978).

The final stage in policy formulation is the actual implementation of the preferred policy strategy. Evaluation procedures here are often known as ex-post or summative evaluations. Typically, they involve identifying how well the implemented strategy has actually satisfied the policy's objectives. Such assessments are not that common in the public policy area (Mitchell, 1979). Jenkins (1978) suggests, for example, that there is often a strong compulsion on the part of many politicians to be unconcerned with a problem once policy has been formulated. It seems that after an issue has evoked a policy response, widespread public interest tends to subside. With little public stimulus, politicians are unlikely to conduct an ex-post evaluation on their own initiative. In addition, many public agencies may hesitate to have decisions scrutinised for fear of mistakes being uncovered.

The two conventional modes of evaluation of public policy, ex-ante and ex-post evaluations, are concerned primarily with implementation strategies and how well they meet given objectives. They have not been used to evaluate the nature of the policy problem or the appropriateness of particular policy objectives. The implicit and unquestioned assumption is that the objectives are worth planning towards (Anderson, 1979). Once we go beyond the narrower interpretations of evaluation, the whole issue of public policy and farmland conservation policies in particular becomes much more difficult. In looking at objectives, we need to recognise that there may be hidden objectives and agenda, which may surface only after the sort of probing analysis that Lowry (1980) undertook for Hawaii's State Land Use Commission. When we undertake evaluations of the impacts and effectiveness of a particular programme, we need to understand that effectiveness (how well are objectives achieved?) is influenced not just by the programme itself but also by external conditions and by other public policy objectives that are being pursued simultaneously (McAllister, 1982). Furthermore, impact assessment can be extended to consider the unforeseen effects of a policy on other groups in society, e.g. a shifting of the tax burden or who really bears the costs of particular programmes.

MOVING BEYOND THE LAND ISSUE. Reviews of governments' reaction to urban encroachment into the City's Countryside reveal that the majority of measures focus on farmland conservation and that this is more developed in North America than elsewhere. But, as FitzSimons (1985, 308) argues, 'policies designed to preserve farmland make little sense unless there are farmers willing and able to produce on those lands'. It has apparently been very difficult for governments to deal with the farm business and farm people issues involved. This has undoubtedly contributed to the low level of information and management provided on a more pro-active basis to farmers.

A concern frequently heard is that in most urban fringe municipalities, the farm population has been outnumbered for a long time and tends to lose much say politically in the community. Many non-farmers, often eager to make a contribution to their new community or, often as eagerly, to affect decisions which are consistent with their self-interest, soon become involved in the local scene. Some become active in politics, while others find that various local boards (e.g. school boards, public utility commissions) are the places they can best make their contribution. As the community structure continues to evolve, many local bodies frequently become dominated by non-farmers who are either unaware of or unconcerned with the problems facing their farming neighbours (FitzSimons, 1985). This situation can even lead to the adoption of by-laws, such as restricting the hours during which certain agricultural procedures can take place, which constrain farming.

Such is the potential tyranny of the democratic system. However, in line with movements towards more sustainable development paths, we can suggest that being politically outnumbered does not mean that the legitimate concerns of minority groups have to be pushed aside. Indeed, there are cases where local areas have taken deliberate steps to ensure that the farm community's perspective is solicited during plan preparation and decision-making generally. The Region of Halton in south-western Ontario, for example, a municipality with a population of more than 250,000 (most of whom are urbanites or rural non-farmers), located about 80 kilometres (50 miles) west of Toronto, established a committee comprising farming representatives (FitzSimons, 1985). The Committee's task is to advise local politicians and bureaucrats on a wide range of issues and to ensure that the farm perspective is taken into account. The formation of the Halton Agricultural Advisory Committee has sent a clear signal to farmers that the local council considers them an important part of the community.

In another example, the City of Brampton, Canada's fastest-growing city in the 1980s, designed a public participation programme specifically to solicit farmers' views during a revision of the municipality's Official Plan. These efforts revealed that farmers were not interested in having land frozen in agriculture, but that they did want development to proceed in an orderly fashion on a predetermined schedule. Phasing development in this way allows farmers to make investment and production decisions in an environment in which one significant element of uncertainty has been removed. At the same time, however, farmers will be able eventually to sell their properties for the best possible price in an area where the pressures are such that limited agricultural futures can be seen in the long term.

On the other hand, the farm/non-farm distinction does not necessarily stand up to scrutiny, as Smit and Flaherty (1981) found when they analysed residents' preferences for different severance policies. Differences in their study area were related to residents' property size, income and length of residence in the area rather than to the farm, rural non-farm and village resident classification.

OTHER FUNCTIONS OF AGRICULTURAL LAND. In the framework presented above (Table 5.4), two other broad categories of goals were identified, those relating to areas where non-agricultural functions of agricultural land are significant and those relating to areas where no agricultural future in the long term is foreseen. Our discussion is very brief because of space limitations.

Where there is no long-term future for the land in agricultural use, because of anticipated conversion to other uses, the key preoccupation is with an orderly and efficient transition. Characteristically, little concern has been expressed for the agricultural communities involved, except in the case of expropriation (see, for example, Bryant, 1975, 1986b). Where land was expropriated for the Charles de Gaulle Airport north-east of Paris in the late 1960s, the farmers won exceptionally handsome financial compensation packages. Owner-occupiers were best off while the tenant farmers who did receive compensation fared less well (Bryant, 1973b). In other situations involving expropriation for the New Towns around Paris in the late 1960s, those farming communities based upon large-scale arable agriculture received compensation through the land prices and other financial packages, while some of the communities based more on intensive farming with smaller acreage, were able to benefit from relocation schemes or in the setting aside of 'permanent' agricultural zones within the New Town area, complete with a renewed agricultural infrastructure (Bryant, 1975; IAURIF, 1978).

Other agricultural concerns in these future non-agricultural zones can also include ensuring agricultural activity is continued as a means of managing the land in the interim, providing adequate notice is given before land is finally withdrawn from production and compensation made for any investment that may have been made since the land was purchased. These are not trivial issues since in some projects it has been twenty years or more since land was purchased for an urban development scheme before any land conversion actually took place. Providing a time schedule for development and keeping to it is no easy matter because progress on large-scale projects can be affected by changing economic and demographic conditions. These long time-frames also mean that while it is understandable that the public authority generally places restrictions on any investments that are relatively fixed and long term, it also may make it difficult to maintain viable agricultural activity under those conditions.

A special case of movement of farmland into another use is conversion to woodlots or forested areas. In the City's Countryside in the European Community, this is likely to occur on some agricultural land removed from farm production as part of the attempts to reduce surplus production. Such wooded areas may provide valuable recreation and ecological functions near cities. Substantial challenges exist, however, in developing policies and programmes that will be effective (Blunden and Curry, 1988).

Non-agricultural functions refer to a wide range of uses for agricultural land, including access to the countryside for recreation, flood control and maintaining the natural, cultural and historic elements of the landscape. The key point here is that these involve ongoing agricultural activity as well, so we are essentially dealing with multiple function and even multiple use of agricultural land. These other functions have become progressively more important.

Physical land use planning and development control have also been used here, e.g. regulations regarding limitations of agricultural structures in a flood plain. In some cases these measures relate to special zones that have been established by some authority above the local and regional level, such as Great Britain's Areas of Outstanding Natural Beauty (Cloke, 1989b), France's *sites classées* and natural reserves (Aitchison, 1989), New Zealand's reserves system (Campbell, 1986) and Australia's conservation reserves, including National Parks (McDonald, 1989). These are basically positive regulatory in orientation. Development control means that in such areas where there is any agriculture farmers may have to seek approval not just for erecting a structure but also for the types and colours of construction materials used (Bryant, 1986b). It is interesting to note that protection measures for environmentally sensitive areas when they are defined 'generously' may have the effect of conserving farmland (cf. the evolving framework of environmentally sensitive areas planning in France (Barcelo and Poitevin, 1988)).

Negotiating easements or acquiring land through trusts (Campbell, 1986; Mason, 1991) or simply through making voluntary agreements with farmland owners and farmers for access to property under certain conditions is also not uncommon. A related method used in the United Kingdom in some areas which has potential for application in areas where it is desired to maintain particular landscapes is to pay farmers to pursue certain practices and not others (O'Riordan, 1986). Similar private law contractual agreements between state and individual farmers have been used in the Netherlands in certain areas designated as management areas (e.g. Grossman, 1987), where a collection of duties and obligations are imposed on both parties (it has still to be widely used, however).

Outright purchase of farmland for these other functions while still maintaining agricultural activity is rare, although sometimes substantial – the Ottawa Green Belt in Canada that has already been mentioned earlier. Another example is the reserve areas for nature and landscape in the Netherlands where the ultimate goal is to end farming through the public purchase of the farmland. Cost-reducing measures to encourage farmers to provide access to their land, for example, can also be arranged in some jurisdictions through tax reductions. There has been some limited experience, e.g. the urban fringe management projects of the British Countryside Commission in the late 1960s and 1970s, in providing funds to help farmers cope with the added pressures that informal non-farm use of their properties may entail. Finally, information and management advice seems again to have substantial potential, but there has probably been more of this for the non-agricultural functions than for adjusting to and taking advantage of the urban fringe for agricultural production and marketing.

AN INTEGRATED APPROACH FOR MULTIPLE FUNCTIONS AND USES? The City's Countryside thus emerges as a broad zone full of complexities and apparent contradictions, related to the many different demands on the land which must be accommodated. Agricultural areas are undoubtedly subjected to many of these pressures, and it is difficult to argue that they should be exempted from assuming some of the responsibility for catering to some of these demands. Reconciliation of these multiple functions is therefore a major challenge.

Rural economies have been transformed substantially over the past half century (Bunce, 1981; Fuguitt et al., 1979; Hansen, 1982). Family farms have been integrated into the agri-business complex (Ball and Heady, 1972), as they have adopted or been replaced by industrial systems (Gregor, 1982; Moore and Dean, 1972; Troughton, 1982a; Whatmore et al., 1986), traditional rural–urban differences are eroding (Glenn and Hill, 1977; Joseph and Smit, 1981; Smit and Flaherty, 1981), non-farm populations are growing and farm populations are shrinking (Wong, 1976), and the functions of many rural villages are changing (Dahms, 1988). Many of these changes have been in response to the various macro forces discussed earlier in the book, including the internationalisation of both capital and trade.

The new structural context, 'the post-industrial society', means that we must devise new and more integrative structures for dealing with the problems of the City's Countryside. No longer can we afford to approach issues in purely sectoral terms. We need to develop integrative structures that allow the various functions that have to be accommodated somehow in the City's Countryside to be considered more comprehensively.

In planning and management approaches and in the minds of some observers, the Green Belt concept offers promise in this regard. More enthusiastically embraced in England than anywhere else, the Green Belt concept may be traced back to Ebenezer Howard (Hall, 1988; Hall et al., 1973;). In response to the overcrowding in urban England resulting from industrialisation, Howard proposed a new form of urban development, the Garden City Concept (Hall, 1988; Sutcliff, 1980). Apart from the design concept of the urban areas themselves (a core and satellite centres), the key relationship of interest to us is how Howard envisaged the link between the cities and the countryside.

Essentially, the urban centres and the 'open' spaces in between would form a single functioning system. The green spaces were to be used for food production, even provide locations for certain urban functions such as institutional uses, and would also serve the recreational demands of the urbanites. The potential was there for the development of a tool that would not only involve integration of rural space within the broader urban region, but also the integration of various functions in a multiple use approach. In reality, emphases have shifted over time and differed between observers (Elson, 1986). While many users of Green Belts concentrate on the recreational and conservation functions, the official view in the United Kingdom emphasises the role of containing urban growth (Munton, 1983b; Strachan, 1974; Thomas, 1970). Indeed, Elson (1986) identifies three purposes served by Green Belts: (1) to check the further growth of a large built-up area; (2) to prevent neighbouring towns from merging; and (3) to preserve the special character of a town – all essentially aimed at controlling and shaping urban development.

An alternative and related concept to the Green Belt is the Green Wedge (cf. Melbourne and Metropolitan Board of Works, 1977). This involves planning for urban growth in an axial pattern, so as to take advantage of presumed efficiences in transportation and servicing, as well as avoid the problems of leapfrogging of development over large swathes of countryside associated with the classic form of the Green Belt. These axial development zones would be separated by wedge-like zones of countryside.

An excellent example of this is the policy that led to the creation of the *zones naturelles d'équilibre* (ZNE) in the Paris region in the mid-1970s (Bryant, 1986b; IAURIF, 1976). Their functions were very similar to those of the English Green Belts: structuring of urban growth, conservation of rural resources, support of recreational and leisure activities, and so forth. Some development was allowed to proceed, and specific zone officers were put in charge of persuading the municipalities in each of the zones to collaborate in various ways and incorporate the philosophy of the ZNE into their land use plans and projects. This has been largely achieved, at least in terms of land use designation, and a fair amount of headway has been made in developing jointly funded projects (municipality, region and central state as well). In fact, the interest in the incentives to participate was sufficiently large for the programme to be extended beyond the original ZNE, and on later regional-scale planning maps the boundaries of the ZNE no longer appear and only a single symbol was used to locate each ZNE.

Currently, the ZNE are no longer even mentioned on the proposed revisions to the master plan but they appear as Natural Areas of Regional Importance (Préfecture de la Région d'Ile-de-France, 1991). Although progress is being made, much remains to be achieved in effective integration of the recreational and conservation-oriented objectives into actual management. The Paris Green Belt in the inner urban fringe is another example of an attempt at an integrated approach. Non-agricultural functions are uppermost in this strategy. Again, while a flexible approach has been adopted, progress has been highly variable between the different geographic sectors (Dubois, 1981).

Goals, public intervention and the geography of the City's Countryside

Images of countryside and policy development

First, it is important to stress that the types of goals and objectives we develop for agriculture as for many other activities are influenced by the underlying images the public, politicians, bureaucrats and professional staff generally have of agriculture.

For many the countryside is a rustic, tranquil place where life moves in concert with nature, at a much slower pace than in the city. However, as Blacksell and Gilg (1981) argue and as the debate in Britain between farmers and conservationists bears testimony (Cox et al., 1985; Munton, 1987), there is an increasing gulf between society's dominant images of rural areas and the realities of modern production. This gulf is important because our images affect the nature of policy and policy affects production.

Consider the conceptual model shown in Figure 5.2. The circle labelled A represents all those agricultural activities which are economically possible in the City's Countryside. Circle B represents those agricultural activities which comply with popularly held images of the countryside. Problems begin with those activities of A that are not part of B. Indeed, it is not difficult to think of situations where public policy formulated in accordance with dominant images denies the existence of some technically and economically feasible agricultural activities in A.

The debate which occurred in New Zealand over the subdivision of land provides an example. Briefly, as we saw in Chapter 2, concern emerged in New Zealand in the early 1970s over the rapid expansion of cities, particularly Auckland and Christchurch. In many areas a minimum size requirement for newly created lots was established – 10 acres was most common in urban fringe areas, although elsewhere other minimum sizes were used (Crawford, 1977; Meister, 1982). After several studies showed that small lots did not have to be less productive than if they were part of a large block (Keng, 1976; Meister and Stewart, 1980; Moran et al., 1980), the idea of an absolute minimum acreage was largely abandoned.

However, in its place, many local areas now apply the 'viable economic unit' test. What this means is that an application for a rural subdivision is granted if it is determined that both of the resulting units are economically

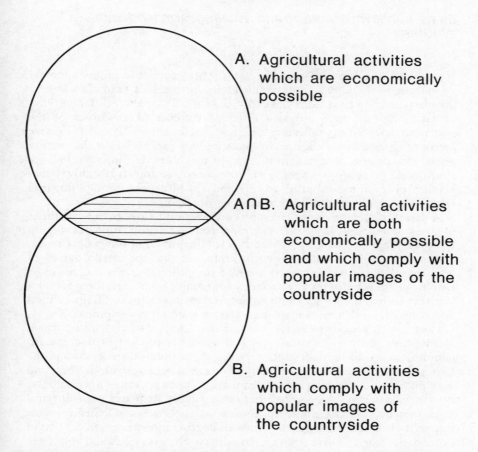

Figure 5.2 Images of the countryside and the realities of modern farm production

viable agricultural operations. Economic viability is generally taken to be the ability of the operation to support an average family. This particular image of agriculture, with the farm providing completely for the needs of the farm household, is only one of many possible farming systems as we have already argued. There is evidence, for example, that multiple job-holding is becoming increasingly frequent in New Zealand (Moran et al., 1989), a situation not accommodated by the image underlying the viable economic unit test as it is generally applied.

Goals, public intervention and evolving farm landscapes: a synthesis

We argued in Chapter 3 that because the structural conditions affecting agriculture in the City's Countryside are more complex than elsewhere, a diverse pattern of agriculture has evolved. Not only are there different types of farming, but there are also different patterns of evolution of the agricultural structure, reflecting different combinations of the formative forces of change in agriculture. Recognising this variety means that setting goals and determining what, if any, forms of public intervention are appropriate must also take these variations into account. It is unlikely that blanket programmes applied to all kinds of situations would function effectively – and yet, this is what happens all too often.

A useful framework is provided by combining the three broad categories of goals used earlier (Table 5.4) with the threefold categorisation of evolving farm landscapes in the City's Countryside (Chapter 3) (Table 5.6) (Bryant, 1984a, 1986b). We should hasten to point out that the categorisation of landscapes (or agricultural areas) is only a simplification, which needs to be empirically developed for any specific geographic region. The three farming landscapes in the City's Countryside referred to are those of agricultural degeneration, agricultural adaptation and agricultural development.

Landscapes of agricultural degeneration are characterised by agriculture which is debilitated. Indeed agriculture is so run-down that its chances of long-term continuation seem limited. Until recently, agricultural degeneration in the City's Countryside dominated many of our images of agriculture there and was linked almost exclusively to urbanisation-based pressures. Increasingly, however, it is being recognised that other factors such as a poor physical environment also contribute to agricultural degeneration. Clearly, it is important to understand the causes of this degeneration in a given context in order to develop effective strategies to achieve whatever goals and objectives might be set for such an area.

Given the difficulties of putting such an area back on its feet, goals for public intervention aimed at boosting or maintaining agricultural production are likely to be rare. If goals for such areas include supporting particular landscapes or providing access for urbanites to the countryside, then very intensive levels of management will be required. On the one hand, the remaining farms would need to be viable, although they could receive payment for maintaining the landscape and providing the access required. In addition, depending upon the causes of the degeneration, it is conceivable that reinforcement of the agricultural structure through some consolidation or *remembrement* (consolidation this time of the fragmented parcels of an individual farm) and through improvement in the farm infrastructure would be required. This type of intensive approach requires more of a persuasive regulatory approach than anything else. It is the sort of approach that will have to be developed in the highly fragmented Paris Green Belt (Dubois, 1981) if this broad strategy is to be successful in tackling many of the smaller agricultural spaces contained within it (Bryant, 1986b). In other cases, the

Table 5.6 Examples of strategies for goal and type of agricultural area combinations

Goals for agricultural areas	Agricultural degeneration	Agricultural adaptation	Agricultural development
Agricultural production goals	A rare combination	Mitigation of negative stresses (e.g. help with fencing) Counselling on marketing strategies Permit certain non-farm enterprises on farms	Little intervention required
Non-agricultural goals (but on-going agricultural activity)	Requires intensive management to create a viable agricultural activity (e.g. farm consolidation, farm infrastructure improvement, management advice) Access and compensation Mitigating measures	Ensure non-agricultural functions do not undermine farming Mitigating measures, e.g. regarding access to farmland and negative impacts Counselling re. adjustment and production/ marketing strategies	As for landscapes of agricultural adaptation
No long-term future for agriculture	Public acquisition of land Compensation Resettlement Adjustment counselling Information on timing and phasing	Public acquisition of land Compensation Resettlement Adjustment counselling Information on timing and phasing	Public acquisition of land Compensation Resettlement Adjustment counselling Information on timing and phasing

goal for such areas will entail ultimate transfer to some other land use. If there was any coercion, e.g. through expropriation, appropriate strategies could include measures aimed at facilitating the orderly withdrawal of the farmers involved, encouraging disinvestment in agriculture, providing life-style and financial counselling, job retraining and resettlement assistance.

Landscapes of agricultural adaptation emerge where farmers can make adjustments in their operations to allow them to cope with or exploit the conditions which prevail in their zones. While there may well be negative impacts created by some forms of development or other factors, a combination of the farmer's abilities and other positive factors in the farming environment in such a zone, e.g. excellent quality farmland and good access to the urban market, allow the adaptation to take place. On the one hand, farmers may be able to cope easily with urban-based irritants that are rather minor (e.g., remembering to remove keys from farm machinery, installing a powerful light in the farmyard, always locking gates), or they can involve major capital outlay (e.g., installing thief-deterrent fences). On the other hand, farmers may engage in strategies designed to exploit or take advantage of local conditions (i.e. positive adaptations (Johnston and Bryant, 1987; Johnston and Bryant, 1989) such as integrating various forms of direct marketing strategies on the farm). The potential is there to reinforce such patterns through appropriate counselling (Ilbery, 1988b).

Where the collective goal for agriculture in such an area entails continued agricultural production, public intervention need not be as intense as in the landscapes of agricultural degeneration. This is principally because some of the farmers have been able to adapt and cope. However, adaptation can be helped and encouraged, or it can be dampened by a lack of flexibility and by the narrowness of some of the images of what constitutes 'good agriculture'. For example, one of the agricultural adaptations encountered on the periphery of Metropolitan Toronto in Johnston's (1989) study was the livery stables and stud farm. Curiously, however, some municipalities in Ontario do not recognise certain equine operations as 'bona fide farms'. This is because equine farms do not fit some people's image of farming, a point that Munton has made for the London Metropolitan Green Belt (Munton, 1983b).

Where non-agricultural functions are seen as primary goals for such areas, e.g. providing access to the countryside for recreation or conserving a nearby wildlife habitat, it is important in developing intervention strategies not to undermine the ongoing agricultural structure, particularly since there will be other stressful factors in such environments (by definition of what constitutes a landscape of agricultural adaptation). In this respect, there are real conflicts between the images created by the industrialised farming and those preferred by recreational users of the countryside (UK Ministry of Agriculture, Fisheries and Food, 1977; Munton, 1983a), especially in the Green Belt approaches. It is particularly difficult because the 'production' (by farmers, landowners) of the values we are concerned with conserving in these contexts (e.g. habitat, landscape value) generally excludes any market compensation to the 'producers' concerned (Crosson, 1989). While this may mean paying attention to the real costs of supporting these other functions for farming, more than anything else it implies the necessity to compromise and to negotiate through a management process the appropriate strategies with the farming population, collectively and individually.

Areas that do not have any long-term agricultural future could undergo interventions as mentioned above to help an orderly and equitable withdrawal of farming.

In *landscapes of agricultural development*, 'normal' agricultural change is dominant, mostly because any negative effects of urbanisation or any other factor are not very intense. Farms in these areas are subject to many of the same external stimuli as those beyond the City's Countryside. Intervention to pursue agricultural production goals is minimal, even non-existent. In the context of such areas being associated either with other non-agricultural functions or with a non-agricultural future (e.g. as in the development of a New Town or other major non-agricultural infrastructural development), then the specific interventions would include many of those already mentioned for these two categories of goals, with the details reflecting the specific set of local circumstances for each farm community.

From a descriptive point of view, it is clearly possible to use this framework (Table 5.6) to describe the set of interventions in any type of agricultural area in a specific City's Countryside. Each cell in the matrix could be associated with a particular set of policy and intervention packages.

The most important conclusion from this discussion is that there is need to recognise the diversity of situations which exists in agriculture in the City's Countryside. From the perspective of public intervention to achieve collective goals for agricultural areas, it is critical for this diversity to be reflected in the policies and intervention strategies that are developed. This requires a level of sensitivity to the needs and situations of individuals and even communities which is rare. It does not mean that positive regulatory approaches are to be discarded – there are many domains where public intervention is necessary to ensure effective, efficient and equitable resource allocation. What we are arguing is that developing policies and interventions without regard for the lives and needs of the individuals concerned contains the seeds for its own destruction. This is even more pertinent given that the nature of many of the issues that have to be resolved in the City's Countryside go far beyond the traditional domain of land use planning and the use of police power. Without co-operation from the farmers and their families, in most countries it is unlikely that rapid progress towards achieving such goals can be made. This is essentially what the enlightened proponents of sustainable development have been arguing, and this is what the experience of the last three decades of public intervention relating to agriculture in the City's Countryside teaches us.

6
The nature of agriculture in the City's Countryside: a synthesis

In this chapter, we offer a set of conclusions on the nature of agriculture in the City's Countryside, based upon the discussion in the first five chapters. We consider first the specificity of agriculture in the City's Countryside by making some comments about its structure compared with that of agriculture in other areas. Then, we synthesise many of the ideas and perspectives discussed to this point by outlining a number of commonly held, but biased, points of view on agriculture in the City's Countryside. Finally, we respond to these by presenting a more realistic and complete picture of agriculture in metropolitan regions.

The specificity of agriculture in the City's Countryside

As a separate and distinct field of scholarly inquiry, the study of agriculture in areas near cities is only about three decades old. Although relatively young in comparison with many other branches of the discipline of geography, the efforts of an international roster of researchers has yielded a significant literature, from which a number of important generalisations can be drawn (Bryant, 1989c) (see Table 6.1 for a summary statement). These generalisations should be set against the backdrop of a more competitive land and labour market and a greater diversity of formative forces in the City's Countryside than elsewhere.

First, metropolitan-centred regions are typically home to a much wider variety of farm types compared with other areas. This even includes the resurgence of mixed farms in some areas. Near-urban zones can be characterised by a patchwork quilt farmscape or mosaic, which in contrast with other areas can seem messy and busy. How many of us have not felt more relaxed by a drive through the comforting monotony of a 'genuine' rural landscape, only to become distraught and agitated by the growing disarray around us as we approach the city? The reasons behind this 'chaos'

Table 6.1 Key characteristics of agriculture in the City's Countryside compared with areas beyond

---

1. A greater variety of farm types, including diversity of specialised farms and greater diversification on farms
2. A greater variety of socio-economic structures of production
   – capitalistic structures
   – part-time and hobby farmers
   – variety of different types of family farms
   – alternative agriculture
   – greater involvement of family members in off-farm work
3. Greater range of capital structures on farms
4. Greater range of types of farm operators
5. Greater development of direct marketing
6. Tendency towards polarisation, bi-modal or greater spread of farms between different business size classes
7. Greater variety of intensities of production and types of intensity (capital-based and labour-based)

---

have been discussed in detail in earlier chapters. Briefly, it is because farms in the City's Countryside are subject not only to the same forces to which other farms are exposed, but also to other forces peculiar to the City's Countryside. On the one hand, the forces of change in the broader society and economy, especially those which have given us the post-industrial society, reach their apogee in the major urban and metropolitan regions. On the other hand, the systems of exchange in which farms function are denser and more varied in this zone, notably the labour and produce marketing systems.

Thus, conditions are created which are favourable both to a variety of socio-economic farming systems as well as a variety of farm types based on enterprise and marketing structures. This is matched therefore by a wide range of capital structures on farms and of farmers and farm managers. Conditions in the City's Countryside present many opportunities for entrepreneurial farmers, and these are epitomised in the many farms with a diversified enterprise structure which have moved into various types of direct selling. The nature of the evolving market for farm produce, and the greater possibility of selling product differentiated by process (e.g. environmentally-friendly produce, organically and biologically produced food) may provide even greater opportunities in the City's Countryside for farmers to participate in lucrative local and regional markets. At the same time, other conditions favour the continued development of industrial farming. While these farms are often run in a business-like manner, they are perhaps more characterised by managerial than entrepreneurial modes of decision-making. Also, there is a range of part-time and hobby farmers. They are a heterogeneous group; some depend very much for their livelihood upon the agricultural revenue, while for others, the agricultural income is incidental to the pleasures of running a farm.

Patterns of tenure, the way in which access to productive land is obtained for farming, is another important distinguishing characteristic. While it is true that many mixed tenure farms (i.e. farms that incorporate both owned and rented land) can be found in almost any rural area, research has shown that farmers in the City's Countryside have taken up the opportunity to rent land with unparalleled zeal. On the other hand, this situation is partly forced on farmers, as there is still a strong preference among farmers for owner-occupation as a means of securing access to the land and to capital.

There is a significant variety of rental agreements. In much of Western Europe, written contracts are fairly common; even without these, the laws governing the rental of farmland provide considerable security to tenants, including fair compensation for tenant investment in the land in the event of a rental agreement being terminated or not renewed. In North America, Australia and New Zealand, all countries without a long history of farmland rental, attitudes towards renting are somewhat different. In general terms, ownership of farmland is still much preferred over rental. Perhaps this is because 'New World' farmers are descended from immigrants who left their homelands because they had little prospect of ever becoming landowners. Indeed, attachment to private property is deeply rooted in many of these countries, land being one of the oldest and most cherished forms of private property.

Notwithstanding this, many farmers in near-urban locations have seen renting of farmland as a viable option. Many farmers there report that they rent land because of the flexibility it affords them in terms of production planning or because it relieves them of the expense of purchasing land. In many jurisdictions, government policy (such as a differential property tax assessment on farmed land) is a powerful incentive for non-farm landowners to make usable parcels of farmland available to farmers on a rental basis. Non-farm landowners clearly benefit from lower tax bills, but it is also reasonable to suggest that many of these people also derive enjoyment from maintaining the agricultural character of their area.

Increased reliance on farmland rental may also be part of a more general trend toward the adoption of industrial methods in agriculture. The renting of farmland parallels the situation in other business sectors where buildings, machinery and many other capital items central to production are leased or rented rather than owned by the firm. One possible outcome is that farmland rental provides yet another conduit through which economic interests external to the farm business can penetrate the farm business.

Farm size distribution, which can be measured in many ways such as farm area, sales volume and total value of assets, provides another dimension in which agriculture in the City's Countryside is distinctive. Some researchers have suggested for instance that a polarisation of agriculture is occurring in the City's Countryside. Rather than the more clearly unimodal distribution with the trend towards farm consolidation that one finds typically in many agricultural areas (and the concomitant marginalisation of other farms), the farm size distribution in the City's Countryside exhibits tendencies towards polarisation or a bi-modal structure or even more complex patterns. This means that the large middle class of farms that dominate the rural countryside are underrepresented in areas near cities. Their place has been taken by a large number of small farms (especially when measured in acreage) and a much smaller cluster of very large farms. The concentration of smaller farms contain many farmers who work off their farms, often on a full-time basis. Others are full-time operations, including intensive horticulture but these are generally in a minority. At the other end of the size scale are found much larger farms. In some cases, they are extremely large, and reflect the extraordinary opportunities for farm consolidation in some urban fringe areas where non-farm ownership of the land is extensive.

It is not surprising that the intensity of farm production also exhibits a tremendous range in the City's Countryside. Generally speaking, it is higher in the City's Countryside than elsewhere, reflecting the greater opportunities for intensive production of fruits and vegetables because of the nearby urban market. However, the range is huge and there is a wide variety of capital structures. On the one hand, there are the most intensive operations of all: horticultural production in greenhouses and feed-lot operations. Although not as intensive, we can add to this end of the spectrum a host of other farms usually serving the urban market, such as vegetable, fruit and poultry production. On the other hand are the specialised arable farms – again intensive compared with many of the farms in the rural peripheries of our economic systems – but definitely more extensive compared with greenhouse horticulture. The less intensive farms also include all those farms, especially in landscapes of agricultural degeneration, where farmers may be mining the soil or where simple operations such as hay-cutting are being undertaken to establish a minimum claim to agricultural use.

This diversity in the intensity with which the land is used partly reflects the differences in the roles that land performs in the City's Countryside. Generally, the role of the land sets agriculture apart from all other economic activities. Land for most agricultural enterprises is one of the critical cornerstones of the farming system. However, land is also important because it provides farmers with a sense of identity, and a basis for distinguishing themselves in society.

However, this is complicated in the City's Countryside. In those areas near cities where much of the farmed land is rented, land is often seen solely as a site for production. Its social role in defining the culture of farming is much less important. And for intensive, highly capitalised farm businesses, such as broiler operations, even if the land is owned by the farmer, then it is still simply a site for production. The declining importance of land as a factor of production is, of course, linked to the broader trend of capital substitution in agriculture, which has accompanied the industrialisation of an important segment of the agricultural sector. Paradoxically, in the extreme forms of extensive production in which technically the land plays a more important role, agricultural land is seen only as a temporary support for farming.

Again paradoxically, while the social and cultural role of farmland in defining an agricultural 'culture' is still important for many of the family farms still present, these roles are perhaps heightened in some of the hobby farms where the farmers and their families do not have any roots in farming.

At this point, it is useful to recall the geographic synthesis suggested in Chapter 5, in which the combination of forces and resulting transformations of farming were discussed from the perspective of three types of evolving agricultural landscapes: degeneration, adaptation and development. While this is a useful way of synthesising patterns, problems and changes, our conclusions so far in this chapter should serve to underline the complexity of the evolving structures. Indeed, it is wise to recognise that this threefold categorisation is only a schematic, that there are potentially many types of evolving landscapes and structures, and that whatever meso pattern or classification scheme we devise, we must acknowledge the real mosaic of situations present in the City's Countryside, as well as the significant variations between and within the regional environments in which the City's Countryside has evolved.

Some commonly held perspectives on agriculture in the City's Countryside

All fields of inquiry can be characterised by a set of commonly held perspectives, the conventional wisdom, which must be questioned and revised periodically. Frequently, the conventional wisdom almost smacks of 'politically correct thinking': is it still a heresy to speak of the agricultural land problem in the City's Countryside being dwarfed by other issues? Dare we raise the notion that part-time farming and hobby farming support important and viable life-styles and that perhaps we should recognise them in our approaches to managing agricultural land?

The myths about agriculture in the City's Countryside revolve around three points:

1. that agriculture there is on a downhill path;
2. that agriculture can be treated as a homogeneous structure; and
3. that most of its problems are linked to the urban environment.

The reality is that agriculture in the City's Countryside is complex in structure (e.g. the different layers of agriculture that partly reflect the many different farming systems and partly the many different 'systems of exchange' in which farms can function there). The reality is also complex in its dynamics. This is also related partly to the different production structures and partly to the variety of forces impinging upon agriculture.

How can we explain the discrepancies that continue to exist between the conventional wisdom and the reality? We suggest they are due to a number of conceptual and methodological biases in much of the research.

One of the most obvious conceptual biases to permeate our thinking in this area is the widespread use of conceptual frameworks that fail to recognise the complex and highly heterogeneous nature of near-urban agriculture and the formative circumstances that underlie the geography of agriculture in the City's Countryside. While it is clearly understood that agriculture in the City's Countryside has responded and continues to respond to urbanisation-based forces, we seem collectively to possess a poor understanding of the ecologically and spatially variable nature of farm-based responses to these forces. For instance, while a host of urban-based stimuli and farm-based responses have been identified, the actual connecting linkages have yet to be the subject of widespread scientific investigation. Furthermore, with some notable exceptions (e.g. Bryant and Greaves, 1978; Best, 1981), little recognition has been given to the manner in which intra-regional differences in the agricultural environment affect the way in which agriculture responds to urbanisation-based influences.

A second bias which is evident in the research literature is the application of conceptual frameworks which do not consider the full range of forces affecting agriculture in the City's Countryside, even as an interpretative backdrop. This bias continues to be present despite the fact that Munton warned of its dangers as early as 1974. Many studies have attempted to explain agricultural change solely on the basis of urban-based influences: classic examples of this in the early literature are the studies of agricultural 'land loss' by Bogue (1956) and Crerar (1963). Such studies underestimate the degree to which agriculture in the City's Countryside also responds to stimuli affecting the general agricultural system.

This bias is perhaps easier to understand in studies which have been based upon the analysis of secondary statistical data, where the evidence, which cannot 'speak for itself' so to speak, can only be interpreted against the particular conceptual framework that has been adopted. This bias has also often been exaggerated in many case studies where farm-level interviews have been undertaken on a local scale, completely within the context of the urban fringe. Agricultural problems are identified certainly – indeed, many studies specifically ask farmers to check off which problems they have encountered in relation to urbanisation. There is nothing wrong with this if there is also ample opportunity for the farmer to respond regarding the whole environment he or she faces. Frequently, this has not been the case. Then, the trap is set! Because the setting is the urban fringe, the classic fallacious assumption is made that the problems identified are linked to the setting alone. More complete conceptual frameworks have been developed which explicitly recognise the complex nature of the many forces affecting agriculture, but they have yet to be widely espoused.

Yet another issue is that many researchers have unquestioningly accepted the notion that interaction between urban development and agriculture results only in negative consequences for agriculture. This has been extended quite readily into the arena of public policy formulation, especially in North America where policies have been developed and adopted to maintain the viability of agriculture in certain regions through the proscription of non-farm development. Admittedly, non-farm development and its associated pressures can under some circumstances undermine the viability of continued agricultural production. However, this general thrust ignores the potential for positive interaction between urban development and agriculture. Outside the immediate active transition zone in the inner urban fringe, farmers for example can benefit from the rental of farmland from non-farm owners of farmland on a long-term basis. Other examples include the stability that can be derived from participation in the non-farm labour market by members of the farm family as well as the adoption of non-traditional marketing strategies for farm produce.

What we are suggesting is that researchers and policy-makers have often accepted partially tested hypotheses and the conventional wisdom as fact. Brown's (1977) study of exurban development in south-western Ontario provides an excellent illustration of this point. He notes that although conflicts between farm and non-farm land uses cannot be measured with any available data, the phenomenon is no less real! The experience of Ontario's Select Committee on Land Drainage in the early 1970s also demonstrates some of the problems with axiomatic acceptance of the conventional wisdom. In a number of briefs submitted to the Committee, problems were described concerning the discontent felt by non-farm residents over farm-generated alterations to natural drainage systems, water table fluctuations and perhaps most important, over the fact that they were contributing to the costs of municipal drains while the farmers were seen to be deriving the principal benefits. Such conditions were believed to be extensive. Because virtually no empirical evidence could be found in the literature to support these claims, a detailed study of seven rural Ontario townships was undertaken (Found et al., 1974). Few problems of the sort outlined in the briefs to the Committee were uncovered, leading the investigators to conclude that while the specific complaints probably had a basis in fact, the generalisations made may have been exaggerated, oversimplified and highly misrepresentative. This simply underlines the difficulties in using the evidence presented at public inquiries, and indeed one might add, through the media, as a means of polling public opinion for the purposes of policy development.

Finally, much of the literature, especially the North American research literature, exhibits a strong, and at times singular focus on land and land use. This is partially explained by the data sources commonly used (e.g. air photos, maps) and by the strong preoccupation with land in the disciplines that have produced many of the researchers in the field: agricultural geography, land use planning, pedology, land economics.

Furthermore, in terms of policy-based research, it could be argued that this bias is also related in North America at least to the fact that planning in the countryside has shied away from dealing with people, households and activities. Yet, we have argued in this book that the interaction between urbanisation and agriculture is as much a people–capital–business phenomenon as a land one. It is easy to ignore these relationships in research that has been based upon secondary data, both census data and mapped land-based data. In neither case is information provided directly on the decision-making unit of analysis which is central to understanding process and change – the combined farm and family unit. Unfortunately, even in much of the interview-based research, the focus either on land or specifically on urban-based problems has been so narrow that opportunities have been missed to elicit information from respondents which might provide alternative insights and explanations. This is no small matter for concern because neglecting these relationships is an oversight that stands to thwart attempts to achieve collective goals for agriculture in the City's Countryside and to provide more realistic and complete explanations for what is happening to agriculture there. This is even more surprising given the increasing attention that has been paid to 'sustainable agricultural production', which entails consideration not just of agricultural land but also of the farming systems.

The concerns

The central preoccupation with agriculture in the City's Countryside in North America remains with agricultural land. Where there is concern over the viability of the agricultural system and the farm business, it is generally because they are seen as contributing to the stability of the land. Somehow, connected with the relatively narrow perspective taken by much of the research, it is assumed that the removal of agricultural land in the City's Countryside reflects the demise of the system. Clearly, in a dynamic system, resources are transferred from one use or function to another. In the context of the City's Countryside, much of the reallocation of the land which affects agriculture is in the inner urban fringe. In these areas, there are certainly examples of premature setting aside of farmland, disinvestment and general degeneration of the farm structure: these areas would be included in the landscapes of agricultural degeneration which we have already provided for in one of the conceptual frameworks introduced in this book.

However, this transfer of agricultural land does not mean that elsewhere in the City's Countryside the agricultural system is collapsing as a result of non-farm development. Non-farm development around the smaller settlements beyond the inner urban fringe, or more scattered development generally – where it still occurs – cannot be expected to lead to the same degeneration of farm business structure compared with those areas where there is massive accretionary development. Furthermore, as we argued earlier, other factors such as the positive influences of urbanisation, other positive aspects of the general agricultural environment in an area and the adaptive and especially entrepreneurial behaviour of some farmers lead to continued agricultural development, either traditional or non-traditional, which may counteract the negative influences at work.

In Western Europe, there has been less preoccupation with the agricultural production capacity lost from the transfer of agricultural land to non-farm uses, undoubtedly partly because of the major, and as yet still unresolved problems arising from surplus agricultural production in the European Common Market. In some instances, as in France, there was more consideration for the communities affected by the land transfers, especially when the resource reallocation involved public decisions and expropriations of land. At the same time, beyond the immediate inner urban fringe, concern centres more on the non-agricultural functions of agricultural land and farming – farming as a landscape support (both 'natural' landscape and the cultural and historic values embedded in the farming landscapes) and as a backdrop for outdoor recreational and leisure-time activities for the nearby urban population. The non-agricultural functions will almost certainly increase in importance even more if the agricultural surplus problem is tackled through extensive setting aside of farmland or extensification of farming methods. Once again, it is important to understand the decision-making unit – the farm and family – in order to clarify the conflicts between farming and these collective non-agricultural goals and the potential for resolution of these conflicts through a management approach to planning for agricultural areas in the City's Countryside.

Is there then no need for concern for agricultural land in the City's Countryside? Not quite! Despite the problem of food surpluses in many West European countries and the restructuring of farming which has led to significant increases in agricultural production, worrisome signals have appeared which suggest that this bountiful situation has occurred only as a consequence of the widespread adoption of industrial farming and over-intensification.

First, this has resulted in a production system that is relatively fragile economically for it is extremely dependent upon external linkages – especially capital and non-farm inputs. This restructuring has hardly been 'natural', even though for decades it was assumed by many to be the normal (and desirable) path for agricultural change. Government after government encouraged this transformation as a means of boosting productivity and keeping food prices low through easy access to the credit necessary to introduce the new capital-intensive technologies (e.g. Dahlberg, 1986; Caldwell, 1988).

Second, as we noted earlier, there has been mounting uneasiness over the impact of this intensification upon the long-term productivity of agricultural land, as well as upon its non-agricultural functions, e.g. support for different types of habitat and the amenity value of farm landscapes. This uneasiness has been expressed widely both in North America and increasingly in Western Europe. While it is difficult to estimate the impact of a potential de-intensification of agriculture on productivity, it does argue for a cautious and conservationist (not preservationist) approach to the conversion of prime agricultural land. At the same time, there are some difficult issues raised by such de-intensification: whether the producer returns would remain 'adequate' and what types of farm structure would evolve. It is interesting to speculate on the role of part-time and hobby farmers here: to the extent that their farms are less intensive, perhaps they contribute both in the short term to resolving the food surpluses and in the long term to maintaining an agricultural resource (Bryant, 1989c).

All of this argues for more comprehensive analysis to understand both the whole farm unit (production plus family relationships) and all the forces impinging on agriculture, operating through the various systems of exchange on different scales. Agriculture in the City's Countryside is not just responding to urban-based forces but also to all the forces that are driving the whole economic and social system. Maybe some of the responses to these broader forces have had a greater negative impact on the economic and social viability of farming in the City's Countryside than the negative urban-based forces.

7
Retrospect and prospect

Agriculture and the city have long enjoyed a special relationship. We have already stressed the importance of the symbiotic relationship between agriculture and development of cities. Agriculture, for example, helped fuel the Industrial Revolution of Europe not only with food but also with labour. In return, urban areas have provided a demand for agricultural commodities and for long have acted as a source for agricultural innovations. Even the use of the countryside for recreation and leisure is not confined to the twentieth century. We need only think of the Romans and their villas, or the seventeenth century weekend and summer 'homes' of the Amsterdam bourgeoisie in the earlier polders in the Netherlands, or the magnificent estates of Europe, which have drawn the monarchs for generations to be reminded that the countryside has long played an important role that goes beyond the production functions associated with the land.

What has changed in the twentieth century is that the interaction between urban and rural areas has become more intense. It now involves a much broader range of activities and groups of people as leisure time, personal mobility and disposable incomes have all increased. The City's Countryside has become more democratised. Meanwhile, the range of 'legitimate interests' in agricultural land has increased and the number of stakeholders in what happens to the City's Countryside has broadened. As it is likely that the underpinnings of sustainable development will gain even more widespread appeal, an important challenge will be to find ways of integrating the interests of these stakeholders into the local and regional level of planning and management which are characteristic of the sustainable development movement.

Recurrent concern about hunger, population growth, resource degradation and resource use conflicts have accompanied the rise of the conservation movement generally since the late nineteenth century. All of this has most certainly influenced the land conservation ethic espoused by many researchers and observers of agriculture in the City's Countryside. The strong link between agricultural production and the biophysical environment is still a critical factor in many people's view. This perspective is, however, as we have already noted in Chapter 5, far more firmly entrenched in North America, especially the United States, than in Western Europe. This is reflected in the greater attention given to agricultural land *preservation* in North America than in Western Europe and to the continued anxiety expressed about the degeneration of agriculture in the City's Countryside there as a result of non-farm development.

It is argued that in Western Europe the issues have been regarded as more complex for a longer time. Part of the difference lies with the North American preoccupation with the land itself, which has meant that until recently, relatively little attention was directed to the other components of the agricultural system. Since any urban expansion is bound to involve the removal of agricultural land from production, a perspective which places such importance on this resource is almost bound to be concerned with the land 'lost'. However, if agriculture is seen in its entirety as involving land, people and capital, then the removal of only one element in some regions is likely to generate less concern. Furthermore, such a broader view of agriculture almost inevitably creates a more comprehensive picture of the whole agricultural system: the entire productive capacity of the system, the range of issues that influence agricultural viability and so forth.

In North America however, perspectives have begun to change. Increasingly, researchers and others are questioning both the necessity and the efficacy of different kinds of farmland conservation measures. Additionally, more attention has been focused upon how the individual farmer and farm operation influences the results of public policy. In the United Kingdom a similar focus has emerged, but more so with respect to the role of the individual farmer in adopting technologies which compromise the integrity of the environment and the non-agricultural functions supported by agricultural land generally. There, efforts have also been directed towards the farm community's attempts to influence public policy aimed at mitigating the negative environmental consequences associated with many modern agricultural practices.

As our focus on the individual has developed, more attention has been given to the real complexities of agricultural structure and change in the City's Countryside. The result is a much more realistic image of agricultural systems in the City's Countryside and an improved understanding of the potential for and limitations of different types of public sector intervention in the City's Countryside. This has occurred side by side with an increasing awareness of the diversity of values associated with farming areas and farming systems in the City's Countryside. A key challenge is therefore to find ways of reconciling these values – agricultural versus non-agricultural, individual versus collective and collective versus collective – so that a minority of the population, the farming community, does not have to shoulder the costs to benefit the majority when that minority is not as well off and well equipped to carry the burden. Furthermore, as local and regional levels of government become more active in the planning of their areas, ways have to be found to incorporate into local area planning the external values and interests that lie beyond (Molnar, 1985).

Agriculture in the City's Countryside today

Understanding agricultural production by placing it in an evolving socio-economic system requires an appreciation of the cornerstones of the farming system: the biophysical resource base, the different markets (e.g. agricultural produce, labour, capital), the relationship between the farm operation and the farm household, and public intervention on different scales. All too often in developing both our explanatory frameworks and various public policy initiatives, we have tended to ignore the complexity of the agricultural system. This is well illustrated in the naïve assumption that is implicit in many programmes of farmland preservation, i.e. that protecting farmland from non-farm development will preserve farming! Indeed, Nelson (1990) argues that the net result of much public intervention in the United States in farmland preservation and conservation has been perverse. This, we suggest, partly reflects a lack of understanding of people's behaviour patterns and the real alternatives with which they are faced.

The quality of the resource base for agriculture in the City's Countryside is undoubtedly very high in many countries. This, as noted earlier, stems from the early links between urban settlement and growth and agriculture in the surrounding area. However, we must recognise that the quality of the biophysical base for agriculture is also variable, even in some of the regions with the most productive land. It must also be accepted that because agricultural technologies vary in their adaptability to different biophysical environments, there will be regional differences in productivity. Moreover, because new technologies and practices are continually being developed, we should also expect spatial differentiation in the relative competitiveness of particular areas within the City's Countryside.

Similarly, there are significant variations in the types of markets for agricultural produce which represent one of the most important sets of systems of exchange that link individual farms directly or indirectly to the consumer. In turn, these contribute to determining the nature of the producing world with which farmers enter into contact. Access to capital also varies from region to region and even within regions, not just as a function of the particular capital circuits present but also as a function of inter-farm differences in socio-economic organisation which influence the perceived need for capital and the eligibility for access to different capital circuits.

We have already stressed the importance of understanding the structure of the decision-making of the individual farmer and farm family. It has improved our comprehension of structural change, but it has also emphasised the inherent complexity of the dynamics involved. Furthermore, it has underlined the important role in any democratic society of the individual farmer and farm family in influencing the nature of public intervention in such areas as agricultural land preservation and conservation and soil conservation programmes, as well as the degree to which policy objectives are achieved. Farmers and their families are certainly individuals with all the inherent variability that entails, even where certain cultural or demographic characteristics, e.g. age profile, appear to imply homogeneity.

With the community dimension of the sustainable development movement beginning to take firm hold, it becomes even more important to understand the individual decision-making level and to find constructive ways to integrate it into general policy. It implies a much more persuasive and management-oriented approach to agricultural issues in the City's Countryside.

Finally, of course, all of this takes place against the backdrop of public intervention in the agricultural system, public intervention in the regulation of land use and the public sector role in moulding the environment within which investment and entrepreneurial decisions are made by farmers and others. The key argument we have tried to make throughout this book has focused on the complexity of patterns of agriculture in the City's Countryside. It is not, however, a complexity that defies interpretation or understanding. Although a difficult task, we believe that by recognising the variety of formative circumstances affecting agriculture, it is possible to create a more balanced and comprehensive picture of the evolving system.

The future: the real issues of urban and agriculture interactions

Our first challenge is to accept the nature of the issues facing agriculture in the City's Countryside without being mesmerised by partial and biased models. There are undeniably areas in the City's Countryside – the landscapes of agricultural degeneration – where negative forces compromise the future of agriculture. Indeed, in some of these areas, urbanisation-based forces may well be the main culprits. In others, however, non-urbanisation forces are significant. The whole range of forces needs to be appreciated to assess realistically the potential public intervention options. We need to question the validity of the images underlying our approaches to planning agricultural areas, e.g. is there a uniformly applicable 'right' form of farming? and are commercially viable farms to be defined as full-time operations?

It is also important to realise that public initiatives affecting agriculture in the City's Countryside do not only confront problems peculiar to agriculture in the City's Countryside. There are in all countries a myriad of other public policies aimed at modifying prices and costs, improving farm income, modifying 'normal' patterns of imports and exports, and influencing production levels. In many ways, these initiatives directly affect the forces that influence farm investment decisions: relative prices, costs and income, and their stability. Consequently, their influence cannot be discounted by those public agencies which are more interested in the 'health' and land use stability of agricultural areas in the City's Countryside. In the European Community, for instance, the evolving policy debate on agricultural support has tremendous implications for agricultural land use in the City's Countryside.

At the same time, it is naïve to think that it is possible to control agricultural land use and guarantee its stability in the City's Countryside simply through influencing those factors that deal with the land and how it is integrated physically into the surrounding environment, e.g. land use zones, road contruction and severances. Land use planning with its focus on physical land use considerations is ill-prepared to deal with the long-term stability of agricultural land use in the City's Countryside. Effective intervention must take into account the economic and social viability of the farm business structure. While regulation of land use may introduce some element of stability into the farmer's planning horizons, it will do nothing for the farmer's future revenue and relatively little for the cost picture.

This does not mean that there is little that can be accomplished. For instance, given the nature of the evolving markets for agricultural produce, with more and more attention being devoted to 'healthy' products and increasing concern being expressed over environmental degradation from a much broader stakeholder base than simply an environmental 'élite', it appears that there are significant opportunities which can be developed in the City's Countryside to link process to product (see, e.g., Moore (1990)). This has already happened to a certain extent with organic or biological agriculture. The potential is there to develop this further, notwithstanding problems that would have to be overcome regarding labelling and quality assurance. Direct selling could circumvent some of these problems, if it is associated with the development of a relationship of trust between farmer and customer. Public intervention is not likely to take the form of regulation however, unless it is to ensure honest advertising. However, there are roles for a management approach in which counselling is offered to farmers on the types of markets to enter and how to market their products (see, e.g., Ilbery's comments (1988b)). This certainly does not fall into the commonly encountered approaches to public intervention in agriculture in the City's Countryside, but it is familiar territory to many of the business counselling programmes that exist in most Western countries to help small business adjust and to take advantage of evolving market opportunities.

This failing of the traditional forms of land use planning is even more dramatically illustrated by the rapidly emerging concerns over the degradation of agricultural soil as a result of modern agricultural technology and farm practices, and the negative impacts of farm technology on the ability of agricultural land to support other non-agricultural functions such as the maintenance of wildlife habitat and landscape amenity values. Once again, it is not our contention that land use planning has no role to play. Rather, a very different mode of operation is required in most countries than use of police power and of land use control and regulation. This requires the development in most cultural contexts of the Western world of a more management-oriented and therefore people-oriented method. We argue this because no matter how easy it is to discuss agriculture in land use terms or even in economic terms, agriculture remains fundamentally an activity undertaken by people. And people in a democracy have a habit of circumventing rules and regulations which they do not perceive as legitimate controls over their (non-criminal) activities. How much more so is this the case when we consider more persuasive and management-oriented approaches.

Another matter that must be addressed is the artificial but quite general separation of rural and non-farm elements in the City's Countryside, a separation that reflects more the underlying ideologies of the bureaucrats and politicians than the realities of incompatibilities between the two elements. Fortunately, the recognition of the economic and social realities of the 'regional city' and the recognition of the multiple values and functions of the agricultural areas which are implicit in the interaction between agricultural and urban land uses in such a settlement system holds out some hope for us to be able to build upon the continued symbiotic relationships between city and countryside. This does not mean that we have to espouse the concept of a 'regional' form of government to look after the management of these regional city systems, although many such regional government structures exist throughout the developed world.

The overriding conclusion to this development of the argument in this book is that it appears inevitable that local and regional forms of government or co-operative structures will assume more and more importance in the management of the community. This movement, related to the increasing sophistication of the population in most developed countries and increased demands for people to have a say in shaping the evolution of 'their own' environment, is part and parcel of the sustainable development movement.

'Think globally, act locally' reflects first the global interdependencies – environmentally, politically and economically – which colour so many of our activities today, and second, the fact that in order to do anything about the negative impacts of development, a major role has to be assumed at the local – and indeed, the individual – level. This does not mean that senior levels of government do not have important roles to play; of course they do. First, there are some aspects of the development of the agricultural system which require negotiation between different national governments. Second, senior governments have an interest in the efficient and effective articulation and implementation of development plans at the local and regional level because these plans may have spin-off effects for a broader population. Partly for this reason, senior governments can still be expected to participate financially in many locally and regionally initiated projects. Third, senior levels of government have major responsibilities for certain standards in, for instance, environmental quality and access to educational and health care.

However, it has proved naïve to expect senior governments to be all things to all people, to be entirely objective in their allocation of public resources and to be efficient and effective in their own undertakings. This also forms part of the backdrop to the increased attention to local and regional involvement in managing the evolution of the local and regional environment.

What needs to be present for this local and regional participation to be effective and equitable? A whole book could be devoted to this topic alone. By way of a conclusion on local and regional involvement in the role of agriculture in the City's Countryside, let us suggest that the first place to start is the development of broadly based and long-term planning for the local or regional community – let us call this a 'strategic planning' framework.

In this process, alternative scenarios for change and development in the community are evaluated, based upon a realistic analysis of the opportunities and constraints associated with achieving the community's vision and objectives. The biggest challenges associated with this are to address the concerns and interests of all the legitimate interests and stakeholders in the community, including those who are not resident there. This requires a substantial degree of openness on the part of the community, and an ability to take externalities into account in local planning processes which extend beyond the community itself. It also requires a willingness to question the conventional wisdom on the definition of a 'true' agricultural area. It is recognised that this does not develop overnight, but it is also recognised that purely top-down approaches to the management of agricultural areas and agricultural land in particular have not functioned particularly well in the long term. An effective partnership needs to be developed between local (and/or regional) levels of government and senior levels.

Land use planning is, of course, related to this strategic planning. However, it is the broader process which can provide the direction for land use planning, AS WELL AS the development of other important means of implementation to achieve the goals of the community. For instance, this might include the creation and implementation of a counselling service for farmers in the marketing area in partnership with a state or provincial or central government agricultural field extension programme. Or it might include facilitating public access to parts of the countryside through the negotiation of access agreements between farmland owners and either the municipality or some other agency or organisation. In strategic planning, which is still in its infancy in the context of local and regional planning, strategic plans *per se* do not carry the weight of the law. However, their strength is that they can be modified as conditions change and they provide a dynamic and coherent reference point for the community against which land use plans, which are not action-oriented, can be measured and modified rationally. In short, in the face of future uncertainty and change involving agriculture and other components of the community, the strategic plan could provide the dynamic blueprint for people in the community to chart the future of agriculture and their community.

Bibliography

Agricultural Institute of Canada (1975), 'A Land Use Policy for Canada', *Agrologist*, 4(4): 22–5.
Aitchison, J. (1989), 'Land-Use Planning in Rural France', in P.J. Cloke (ed.), *Rural Land-Use Planning in Developed Nations*, London, Unwin Hyman, 76–103.
Alduy, J.-P. (1983), 'Quarante ans de planification en région d'Ile-de-France', Cahiers de l'Institut d'Aménagement et d'Urbanisme de l'Ile-de-France, 70: 11–67.
Allen, P., and Van Dusen, D. (eds) (1988), *Global Perspectives on Agroecology and Sustainable Agricultural Systems*, Santa Cruz, California, University of California.
Allison, R. (1984), 'Agricultural Land Use and Farming Systems in Southern Ontario', Plymouth, England, College of St Mark and St John, Canadian Studies Geography Project for Sixth Forms and Colleges.
Anderson, J.E. (1979), *Public Policy Making*, New York, Holt, Rinehart and Winston.
Archer, R.W. (1973), 'Land Speculation and Scattered Development: Failures in the Urban Fringe Land Market', *Urban Studies*, 10: 367–72.
Archer, R.W. (1978), *Policy and Research Issues in Subdivision for Rural Residences: Hobby Farms and Rural Retreats*, Canberra, Department of Urban and Regional Development – AGPS.
AREEAR (Atelier Régional d'Etudes Economiques et d'Aménagement Rural) (1976), *L'agriculture spécialisée en Ile-de-France*, Paris, France, Ministry of Agriculture.
Bairoch, P. (1988), *Cities and Economic Development: From the Dawn of History to the Present*, London, Mansell.
Baker, S. (1989), 'A Study of Erosion Potential on Rented and Owned Agricultural Land in Norwich Township, Oxford County, Ontario', Waterloo, Department of Geography, University of Waterloo, unpublished Senior Honours Essay.
Ball, A.G., and Heady, E.O. (eds) (1972), *Size, Structure, and Future of Farms*, Ames, The Iowa State University Press.
Barcelo, S. and Poitevin, J. (1988), *Les espaces naturels sensibles des Yvelines*, Paris, Institut d'Aménagement et d'Urbanisme de la Région d'Ile-de-France.
Barkema, A.D. (1987), 'Farmland Values: The Rise, the Fall, the Future', *Economic Review*, Federal Reserve Bank of Kansas City, April: 19–35.
Barlowe, R. (1986), *Land Resource Economics: The Economics of Real Estate* (4th edn), Englewood Cliffs, Prentice-Hall.
Barron, J.C. and Thompson, J.W. (1973), *Impacts of Open Space Taxation in Washington*, Pullman, Washington, Washington Agricultural Experimental Station, Washington State University, Bulletin 772.
Baschung, M. (1987), '15 ans d'aménagement du territoire au niveau fédéral', *Aménagement du Territoire: Bulletin d'Information*, 1/2: 3–15.
Bately, R. (1983), *Power Through Bureaucracy: Urban Political Analysis in Brazil*, Aldershot, Gower.
Bauer, G., and Roux, J.-M. (1976), *La rurbanisation*, Paris, Editions du Seuil.
Beale, C.L. (1974), 'Rural Development: Population and Settlement Prospects', *Journal of Soil and Water Conservation*, 29(1): 23–7.
Bell, D. (1973), *The Coming of Post-Industrial Society*, New York, Basic Books.
Benevolo, L. (1980), *The History of the City*, London, Scolar Press.

Benmoussa, M., Favrichon, V., Ginot, V. and Labarre, T. (1984), *Activité équestre et propositions d'aménagement dans le Parc Naturel Régional de la Haute Vallée de Chevreuse*, Paris, Ecole Nationale du Génie Rural, des Eaux et des Forêts, Mémoire d'études.

Berry, D. (1978), 'Effects of Urbanization on Agricultural Activities', *Growth and Change*, 9: 2–8.

Berry, D. (1979), 'The Sensitivity of Dairying to Urbanization: A Study of Northeastern Illinois', *The Professional Geographer*, 31: 170–6.

Berry, E., Leonardo, E. and Bieri, K. (1976), *The Farmers' Response to Urbanization: A Study of the Middle Atlantic States*, Philadelphia, Pennsylvania, Regional Science Research Institute, Discussion Paper 92.

Berteloot, F.Y. (1972), *Les plans d'occupation des sols*, Nantes France, Chambre d'Agriculture de Loire Atlantique.

Best, R.H. (1978), 'Myth and Reality in the Growth of Urban Land', in A. Rogers (ed.), *Urban Growth, Farmland Losses and Planning*, London, Institute of British Geographers, 2–15.

Best, R.H. (1981), *Land Use and Living Space*, London, Methuen.

Best, R.H. and Coppock, J.T. (1962), *The Changing Use of Land in Britain*, London, Faber and Faber.

Biancale, M. (1982), *Les côteaux de Chambourcy à Orgeval*, Paris, France, Institut d'Aménagement et d'Urbanisme de la Région d'Ile-de-France.

Blacksell, M. and Gilg, A. (1981), *The Countryside: Planning and Change*, London, George Allen and Unwin.

Blair, A.M. (1980), 'Urban Influences on Farming in Essex', *Geoforum*, 11: 371–84.

Blunden, J. and Curry, N. (1988), *A Future for Our Countryside*, Oxford, Basil Blackwell Ltd.

Boal, F.W. (1970), 'Urban Growth and Land Value Patterns', *Professional Geographer*, 22: 79–82.

Bogue, D.J. (1956), *Metropolitan Growth and the Conversion of Land to Non-Agricultural Uses*, Ohio, Scripps Foundation.

Bollman, R.D. (1988), 'Revenu des familles agricoles, Canada et Québec, 1957–1986', *Recherches Sociographiques*, 29: 265–81.

Boserup, E. (1965), *The Conditions of Agricultural Growth: The Economics of Agrarian Change under Population Pressure*, London, George Allen and Unwin.

Bowler, I.R. (1981a), 'Self Service down on the Farm', *Geography*, 66: 147–50.

Bowler, I.R. (1981b), 'Some Characteristics of an Innovative Form of Agricultural Marketing', *Area*, 13: 307–14.

Braden, J.B. (1982), 'Some Emerging Rights in Agricultural Land', *American Journal of Agricultural Economics*, 64(1): 19–27.

Braybrook, D., and Lindblom, C. (1963), *A Strategy of Decision*, London, The Free Press.

Brklacich, M.J. (1989), 'A Framework for Evaluating the Sustainability of Food Production Systems in a Changing Environment', Waterloo, Canada, Department of Geography, University of Waterloo, unpublished PhD thesis.

Brklacich, M.J., Bryant, C.R. and Smit, B.E. (1991), 'A Review and Appraisal of the Concept of Sustainable Food Production', *Environmental Management*, 15(1): 1–14.

Brklacich, M.J., Smit, B.E., Bryant, C.R. and Dumanski, J. (1989), 'Impacts of Environmental Change on Sustainable Food Production in Southwest Ontario, Canada', *Land Degradation and Rehabilitation*, 1: 291–303.

Brown, L.R. (1981), 'World Population Growth, Soil Erosion, and Food Security', *Science*, 214: 995–1002.

Brown, R. (1977), *Exurban Development in Southwestern Ontario – Forms and Effects upon the Region's Rural Resources*, Toronto, Ontario Ministry of Treasury, Economics and Intergovernmental Affairs.

Brunet, P. (1960), 'Structure agraire et économie rurale des plateaux tertiaires entre la Seine et l'Oise', Caen, Université de Caen, Doctorat ès-Lettres.

Brunet, Y. (1980), 'L'exode urbain, essai de classification de la population des Cantons de L'Est', *The Canadian Geographer*, 24(4): 385–405.

Bryant, C.R. (1970), 'Urbanization and Agricultural Change Since 1945: A Case Study From the Paris Region', London, London School of Economics and Political Science, Universtiy of London, unpublished PhD thesis.

Bryant, C.R. (1973a), 'L'agriculture face à la croissance metropolitaine: le cas des exploitations fruitiére de Groslay et Deuil-la-Barre dans la grande banlieue nord de Paris', *Economie Rurale*, 98: 35–55.

Bryant, C.R. (1973b), 'L'agriculture face à la croissance metropolitaine: le cas des exploitations de grande culture expropriées par l'emprise de l'Aéroport Paris-Nord', *Economie Rurale*, 95: 23–35.

Bryant, C.R. (1974), 'The Anticipation of Urban Expansion: Some Implications for Agricultural Land Use Practices and Land Use Zoning', *Geographia Polonica*, 28: 93–115.

Bryant, C.R. (1975), 'Metropolitan Development and Agriculture: The SAFER de l'Ile-de-France', *Land Economics*, 51(2): 158–63.

Bryant, C.R. (1976), *Farm-Generated Determinants of Land Use Change in the Rural–Urban Fringe in Canada, 1961–1975*, Ottawa, Lands Directorate, Environment Canada, Technical Report.

Bryant, C.R. (1981), 'Agriculture in an Urbanizing Environment: A Case Study from the Paris Region, 1968 to 1975', *The Canadian Geographer*, 21(1): 27–45.

Bryant, C.R. (1982), *The Rural Real Estate Market: An Analysis of Geographic Patterns of Structure and Change within an Urban Fringe Environment*, Waterloo, Canada, Department of Geography, University of Waterloo, Publication 18.

Bryant, C.R. (1984a), 'The Recent Evolution of Farming Landscapes in Urban-Centred Regions', *Landscape Planning*, 11: 307–26.

Bryant, C.R. (1984b), 'Agriculture in the Urban Fringe: A Systems Perspective', *Rural Systems*, II: 1–15.

Bryant, C.R. (1986a), 'Agriculture and Urban Development', in M. Pacione (ed.), *Progress in Agricultural Geography*, Beckenham, UK, Croom Helm, 167–94.

Bryant, C.R. (1986b), 'Farmland Conservation and Farming Landscapes in Urban-Centred Regions: The Case of the Ile-de-France Region', *Landscape and Urban Planning*, 13: 251–76.

Bryant, C.R. (1988), 'Economic Activities in the Urban Field', in P.M. Coppack, L.H. Russwurm and C.R. Bryant (eds), *The Urban Field: Essays on Canadian Urban Process and Form III*, Waterloo, Canada, Department of Geography, University of Waterloo, Publication 30, 57–79.

Bryant, C.R. (1989a), 'Rural Land-Use Planning in Canada', in P.J. Cloke (ed.), *Rural Land-Use Planning in Developed Nations*, London, Unwin Hyman, Chapter 8, 178–206.

Bryant, C.R. (1989b), 'Entrepreneurs in the Rural Environment', *Journal of Rural Studies*, 5: 337–48.

Bryant, C.R. (1989c), 'L'agriculture péri-urbaine au Canada: dégénérescence ou nouvelle dynamique', *Annales de Géographie*, 548: 403–20.

Bryant, C.R. (1990), 'Les grandes régions métropolitaines au Canada: les choix du XXIe siècle', *Etudes Canadiennes*, 29: 7–18.

Bryant, C.R. and Coppack, P.M. (1991), 'The City's Countryside', in T. Bunting and P. Filion (ed.), *Canadian Cities in Transition*, Toronto, Oxford University Press, 209–38.

Bryant, C.R. and Greaves, S.M. (1978), 'The Importance of Regional Variation in the Analysis of Urbanisation–Agriculture Interactions', *Cahiers de Géographie du Québec*, 22(57): 329–48.

Bryant, C.R., et al. (Johnston, T.R.R., LeDrew. E.F., Zhang, Q., Marois, C., DesLauriers, P., and Cavayas, F.) (1989a), 'Agricultural Intensification and Change near Toronto and Montréal', Baltimore, paper presented at the annual conference of the Association of American Geographers, 20 March.

Bryant, C.R., et al. (LeDrew, E.F., Marois, C. and Cavayas, F.) (eds) (1989b), *Remote Sensing and Methodologies of Land Use Change Analysis*, Waterloo, Canada, Department of Geography, University of Waterloo, Occasional Paper 6.

Bryant, C.R., Marois, C., Laurendeau, S. and Deslauriers, P. (1991), 'Behavioural Considerations in the Interpretation of Farm Adaptation and Change in the Toronto and Montréal Fringes, Kingston', paper presented at the annual meetings of the Canadian Association of Geographers, June 1991.

Bryant, C.R. and Russwurm, L.H. (1979), 'The Impact of Nonagricultural Development on Agriculture: A Synthesis', *Plan Canada*, 19(2): 122–39.

Bryant, C.R. and Russwurm, L.H. (1982), 'North American Farmland Protection Strategies in Retrospect', *GeoJournal*, 6(6): 501–11.

Bryant, C.R., Russwurm, L.H. and McLellan, A.G. (1982), *The City's Countryside: Land and its Management in the Rural–Urban Fringe*, London, Longman.

Bryant, C.R., Russwurm, L.H., and Wong, S.-Y. (1981), 'Census Farmland Change in Canadian Urban Fields, 1941–1976', *Ontario Geography*, 18: 7–23.

Bryant, C.R., Russwurm, L.H., and Wong, S.-Y. (1984), 'Agriculture in the Urban Field: An

Appreciation', in M.F.Bunce and M.J. Troughton (eds), *The Pressures of Change in Rural Canada*, Toronto, Department of Geography, York University, Atkinson College, Geographical Monograph 14, Chapter 2, 12–33.

Bryant, C.R., and Wong, S.-Y. (1986), 'Changing Agricultural Structures around Canadian Metropolitan Centres, 1941 to 1976: A Typological Approach', in V.R. Singh and N.K. Singh (eds), *Perspectives in Agricultural Typology*, Varanasi, India, Star Distributors, 174–96.

Bryant, R.W.G. (1972), *Land: Private Property, Public Control*, Montreal, Harvest House.

Bryant, W.R., and Conklin, H.E. (1975), 'New Farmland Preservation Programs in New York', *Journal of the American Institute of Planners*, 41(6): 390–6.

Buckland, J.G. (1987), 'The History and Use of Purchase of Development Rights in the United States', *Landscape and Urban Planning*, 14(3): 237–52.

Bunce, M. (1981), *Rural Settlement in an Urban World*, London, Croom Helm.

Bunce, M.F. (1984), 'Agricultural Land as a Real Estate Commodity: Implications for Farmland Preservation in the North American Urban Fringe', *Landscape Planning*, 12: 177–92.

Bunce, M.F. (1991), 'Local Planning and the Role of Rural Land in Metropolitan Regions: The Example of the Toronto Area', in G.M.R.A. van Oort, L.M. van den Berg, J.G. Groenendijk and A.H.H.M. Kempers (eds), *Limits to Rural Land Use*, Wageningen, The Netherlands, Centre for Agricultural Publishing and Documentation (Pudoc), 113–22.

Buttel, F. (1982), 'The Political Economy of Agriculture in Advanced Industrial Societies', *Current Perspectives in Social Theory*, 3: 27–55.

Buttel, F. and Gillespie, G. (1984), 'The Sexual Division of Farm Household Labour: An Exploratory Study of the Structure on On-Farm and Off-Farm Labour Allocation Among Farm Men and Women', *Rural Sociology*, 49(2): 183–209.

Caldwell, G. (1988), 'La surcapitalisation de l'agriculture québécoise et l'idéologie de l'entreprise', *Recherches Sociographiques*, 29(2–3): 349–71.

Campbell, G.H. (1986), 'Planning Strategies for Nature Conservation', Wellington, New Zealand, The National Trust Lecture 2, The Queen Elizabeth II National Trust.

Campbell, R.R. and Johnson, D.M. (1976), 'Propositions on Counterstream Migration', *Rural Sociology*, 41(1): 127–45.

Carman, H.F. (1977), 'California Landowners' Adoption of a Use-Value Assessment Program', *Land Economics*, 53(3): 275–87.

Chapman, G.R., Smit, B. and Smith, W.R. (1984), 'Flexibility and Criticality in Resources Use', *Geographical Analysis*, 16(1): 52–64.

Chassagne, M.-E. (1980), 'Quelle agriculture dans une société post-industrielle?' *Politique Aujourd'hui*, 1.

Chisholm, M. (1962), *Rural Settlement and Land Use: An Essay in Location*, London, Hutchinson.

Chisholm, M. (1966), *Geography and Economics*, London, G. Bell.

Chung, J.H. (1972), *Land Market and Land Speculation*, Ottawa, Central Mortgage and Housing Corporation.

Clarke, G.R., and Simpson, I.G. (1959), 'A Theoretical Approach to Profit Maximisation Problems in Farm Management', *Journal of Agricultural Economics*, 13: 250–1.

Clawson, M. and Hall, P. (1973), *Planning and Urban Growth: An Anglo-American Comparison*, Baltimore, The Johns Hopkins Press.

Clemenson, H.A. (1985), 'Farmland Rental in Ontario: A Growing Trend', in T. Fuller (ed.), *Farming and the Rural Community in Ontario: An Introduction*, Toronto, Foundation for Rural Living, 181–96.

Cloke, P.J. (ed.) (1989a), *Rural Land-Use Planning in Developed Nations*, London, Unwin Hyman.

Cloke, P.J. (1989b), 'Land-Use Planning in Rural Britain', in P.J. Cloke (ed.), *Rural Land-Use Planning in Developed Nations*, London, Unwin Hyman, 18–46.

Cocklin, C.R. (1981), 'Future Urban Expansion and Implications for Agricultural Land in Ontario', Guelph, Ontario, Department of Geography, University of Guelph, unpublished MA thesis.

Codrington, S.B. (1979), 'The Milk Zone on New South Wales', *Geography Bulletin*, 11(4): 122–32.

COG (Canadian Organic Growers) (1989) *Organic Agriculture Directory*, Montreal, Les Editions Humus Inc.

Conaway, K.M. (1990), 'Carrol County Maryland: Balancing Growth with Agriculture', in R.

Corbett (ed.), *Protecting our Common Future: Conflict Resolution within the Farming Community*, Sackville, New Brunswick, Rural and Small Town Research and Studies Programme, Mount Allison University, 55–62.

Conklin, H.E. (1980), *Preserving Agriculture in an Urban Region*, Ithaca, New York, New York's Food and Life Sciences Bulletin, Report 8.

Coote, D.R., Dumanski, J. and Ramsey, J.F. (1981), *An Assessment of the Degradation of Agricultural Land in Canada*, Ottawa, Land Resource Research Institute, Agriculture Canada, Contribution 18.

Coppack, P.M. (1985), 'A Stage Model of Central Place Dynamics in Toronto's Urban Field', *East Lakes Geographer*, 20: 1–13.

Coppack, P.M., Russwurm, L.H. and Bryant, C.R. (eds) (1988), *The Urban Field: Essays on Canadian Urban Process and Form III*, Waterloo, Canada, Department of Geography, University of Waterloo, Publication 30.

Corbett, R. (1990), *Protecting our Common Future: Conflict Resolution within the Farming Community*, Sackville, New Brunswick, Rural and Small Town Research and Studies Programme, Mount Allison University.

Cosgrove, D.E. (1984), *Social Formation and Symbolic Landscape*, London, Croom Helm.

Coughlin, R.E. et al. (1977), *Saving the Garden: The Preservation of Farmland and Other Environmentally Valuable Land*, Philadelphia, Regional Science Research Institute.

Coughlin, R.E. (1979), 'Agricultural Land Conversion in the Urban Fringe', in M. Schepf (ed.), *Farmland, Food and the Future*, Ankeny, Soil Conservation Society of America, 29–48.

Coughlin, R.E., and Keene, J.C. (eds) (1981), *The Protection of Farmland: A Reference Guidebook for State and Local Government*, Washington, United States Government Printing Office.

Countryside Commission (1976), *The Bollin Valley: A Study of Land Management in the Urban Fringe*, London, HMSO, Countryside Commission CCP97.

Cox, G., Lowe, P. and Winter, M. (1985), 'Caught in the Act: The Agricultural Lobby and the Conservation Debate', *Ecos*, 6(1): 18–23.

Cox, K.R. (1973), *Conflict, Power and Politics in the City: A Geographical View*, New York, McGraw-Hill.

Coye, M.J. (1986), 'The Health Effects of Agricultural Production', in K.A. Dahlberg (ed.), *New Directions for Agriculture and Agricultural Research: Neglected Dimensions and Emerging Alternatives*, Totowa, NJ, Rowman and Allanheld, 165–98.

Crabb, P. (1984), 'Agricultural Change and Agricultural Systems', *Geography Bulletin*, 15(4): 261–74.

Crawford, P. (1977), *Small Rural Lots: A Survey and Commentary on Small Rural Lots and Rural Planning in Taupo County*, Taupo, Taupo County Council.

Crerar, A.D. (1963), 'The Loss of Farmland in the Growth of the Metropolitan Regions of Canada', in *Resources for Tomorrow* (Background papers: Supplementary Volume), Ottawa, The Queen's Printer, 181–96.

Crewson, D., and Reeds, L.G. (1982), 'Loss of Farmland in South-Central Ontario from 1951 to 1971', *The Canadian Geographer*, 26(4): 355–60.

Crickmer, R. (1976), 'The Role of Physical Factors in the Processes of Farmland Abandonment: The Case of Nova Scotia, 1953–1974', Waterloo, University of Waterloo, Department of Geography, unpublished MA thesis.

Crosson, P. (1989), 'The Long Term Adequacy of Agricultural Land: A Different Perspective', Guelph, Ontario, University of Guelph, Department of Agricultural Economics and Business, J.S. McLean Memorial Lecture.

Cruikshank, A.B. (1982), *Where Town Meets Country: Problems in Peri-Urban Areas in Scotland*, Aberdeen, Aberdeen University Press.

Dahlberg, K.A. (ed.) (1986), *New Directions for Agriculture and Agricultural Research: Neglected Dimensions and Emerging Alternatives*, Totowa, New Jersey, Rowman and Allanheld.

Dahlberg, K.A. (1988), 'Ethical and Value Issues in International Agricultural Research', *Agriculture and Human Values*, 5(1 & 2): 101–11.

Dahlberg, K.A. (1989), 'Sustainable Agriculture in an Uncertain Future', Palmerston North, New Zealand, seminar presented at Massey University, 23 June.

Dahms, F.A. (1988), *The Heart of the Country: From the Great Lakes to the Atlantic Coast – Rediscovering the Towns and Countryside of Canada*, Toronto, Deneau.

Daniels, T.L., Lapping, M.B. and Keller, J.W. (1989), 'Rural Planning in the United States:

Fragmentation, Conflict and Slow Progress', in P. Cloke (ed.), *Rural Land-Use Planning in Developed Nations*, London, Unwin Hyman, 152–77.

Daniels, T.L. and Reed, D.E. (1988), 'Agricultural Zoning in a Metroplitan County: An Evaluation of the Black Hawk County, Iowa, Program', *Landscape and Urban Planning*, 16: 303–10.

Davidson, J. and Wibberley, G.P. (1977), *Planning and the Rural Environment*, Oxford, England, Pergamon.

Davies, R.D. (1972), *Preserving Agricultural and Open-Space Lands: Legislature Policy-Making in California*, Davis, Institute of Governmental Affairs, University of California, Environmental Quality Series 10.

Dawson, A.H. (1982), 'Unused Land on the Urban Fringe in Scotland', in A.B. Cruickshank (ed.), *Where Town Meets Country: Problems in Peri-Urban Areas in Scotland*, Aberdeen, Aberdeen University Press, 12–26.

Dawson, A.H. (1984), *The Land Problem in the Developed Economy*, London, Croom Helm.

Dawson, A.H. (1987), 'Farm Size and Land Use in the Urban Fringe of Scotland', in C. Cocklin, B.E. Smit and T.R.R. Johnston, (eds), *Demands on Rural Lands: Planning for Resource Use*, Boulder, Westview Press, 277–88.

Debailleul, G. (1988), 'Zonage et agriculture québécoise dans les années 70', *Recherches Sociographiques*, 29(2–3): 397–416.

DeJong, G.F. and Humphrey, C.R. (1976), 'Selected Characteristics of Metropolitan-to-Nonmetropolitan Area Migrants', *Rural Sociology*, 41(4): 526–38.

Denman, D.R. and Prodano, S. (1972), *Land Use: An Introduction to Proprietary Land Use Analysis*, London, George Allen and Unwin.

Department of Agriculture and Fisheries for Scotland (1982), 'Agriculture on the Urban Edge', in A.B. Cruickshank (ed.), *Where Town Meets Country: Problems in Peri-Urban Areas in Scotland*, Aberdeen, Aberdeen University Press, 61–6.

Department of Environment and Planning (Sydney) (1984), *Sydney Region: North West Sector, Regional Environmental Study*, Volume 1, Sydney, Department of Environment and Planning.

DesLauriers, P., Marois, C. and Bryant, C.R. (1990), 'Diversification Strategies in the Rural–Urban Fringe South of Montréal', Toronto, paper presented at the annual conference of the Association of American Geographers, April.

Detwiler, P.M. (1980), *Rejecting Centralism: An Argument for Improving California's Existing Planning*, Sacramento, California Office of Planning and Research, mimeo.

Diemer, H.L. (1974), *Parkland County Country Residential Survey*, Edmonton, Alberta Land Use Forum, Report 4a.

Dubois, J.-L. (1981), *Etude préliminaire d'ensemble de la ceinture verte*, Paris, Institut d'Aménagement et d'Urbanisme de la Région d'Ile-de-France.

Dyer, A.G., Smit, B.E., Brklacich, M. and Rodd, R.S. (1982), 'A Land Evaluation Model (LEM): Concept, Design and Application', *Computers, Environment and Urban Systems*, 7(4): 367–76.

Edwards, A.M. (1969), 'Land Requirements for UK Agriculture by the Year 2000: A Preliminary Statement', *Town and Country Planning*, 37: 108–15.

Edwards, C.J.W. (1978), 'The Effects of Changing Farm-Size upon Levels of Farm Fragmentation', *Journal of Agricultural Economics*, 29, (1): 143–54.

Elson, M.F. (1986), *Green Belts: Conflict Mediation in the Urban Fringe*, London, Heinemann.

Embleton, C. and Coppock, J.T. (eds) (1968), *Land Use and Resources: Studies in Applied Geography*, London, Institute of British Geographers, Special Publication 1.

Environment Canada (1989), 'Urbanization of Rural Land in Canada, 1981–86: A State of the Environment Fact Sheet', Ottawa, Sustainable Development Branch, Environment Canada, SOE Fact Sheet 89–1.

Errington, A. (1986), *The Changing Structure of the Agricultural and Horticultural Workforce*, Reading, Agricultural Manpower Society.

Ervin, D.E., Fitch, J.B., Gofwin, R.K., Shepard, W.B. and Stoevener, H.H. (eds) (1977), *Land Use Control: Evaluating Economic and Political Effects*, Cambridge, Ballinger.

Esseks, J.D. (1978), 'The Politics of Farmland Preservation', in D.F. Hadwinger and W.P. Browne (eds), *The New Politics of Food*, Lexington, MA, Lexington Books, D.C. Heath, 199–216.

Etzioni, A. (1973), 'Mixed-Scanning: A Third Approach to Decision-making', in A. Faludi (ed.), *A Reader in Planning Theory*, Oxford, Pergamon Press, 217–29.
Falcon, W., Kurien, C., Monckeberg, F., Okeyo, A., Olayide, S., Rabar, F. and Tims, W. (1987), 'The World Food and Hunger Problem: Changing Perspectives and Possibilities, 1974–1984', in Faludi, A. (1973a), *Planning Theory*, Oxford, Pergamon Press.
Faludi, A. (1973b), *A Reader in Planning Theory*, Oxford, Pergamon Press.
Faludi, A. (1987), *A Decision-Centred View of Environmental Planning*, Oxford, Pergamon Press.
Feick, R.D. (1991), 'The Design of an Exploratory Model for Planning School Facility Systems', Waterloo, Department of Geography, University of Waterloo, unpublished MA thesis.
Fielding, J.A. (1979), 'Farmland Rental in an Urbanizing Environment: The Fringes of Kitchener, Waterloo and Cambridge', 1971–78, Waterloo, Department of Geography, University of Waterloo, unpublished MA thesis.
Firey, W. (1946), 'Ecological Considerations in Planning for Rurban Fringes', *American Sociological Review*, 11: 411–23.
FitzSimons, J. (1985), 'Urban Growth: Its Impact Upon Farming and Rural Communities', in A.M. Fuller (ed.), *Farming and the Rural Community in Ontario: An Introduction*, Toronto, Foundation for Rural Living, 297–313.
Flaherty, M.S., Chapman, G.R. and Smit, B.E. (1988), 'Land-Use Criticality Measures Based on an Interior Point in a Convex Polytope', *Environment and Planning B: Planning and Design*, 15(1): 37–46.
Flaherty, M., and Smit, B.E. (1982), 'An Assessment of Land Classification Techniques in Planning for Agricultural Use', *Journal of Environmental Management*, 15(4): 323–32.
Flaherty, M.S., Smit, B.E., Webber, M.J. and Reeds, L.G. (1987), 'Agricultural Land-Use Flexibility: The Concept and Some Measures', *Resource Management and Optimization*, 6(2): 29–51.
Forkenbrock, D.J. and Fisher, P.A. (1980), *Tax Incentive Options to Slow Farmland Conversion in Iowa*, Ames, Iowa, University of Iowa, Legislative Environmental Advisory Group Series, Report 70–30.
Found, W.C. (1971), *A Theoretical Approach to Rural Land-Use Patterns*, New York, St Martins Press.
Found, W.C., Hill, A.R. and Spence, E.S. (1974), *Economic and Environmental Impacts of Land Drainage in Ontario*, Toronto, Department of Geography, York University/Atkinson College, Geographical Monograph 6.
Frankena, M.W. and Scheffman, D.T. (1980), *Economic Analysis of Provincial Land Use Policies in Ontario*, Toronto, Ontario Economic Council.
Frederic, P.B. (1991), 'Public Policy and Land Development: The Maine Land Use Regulation Commission', *Land Use Policy*, January: 50–62.
Friedmann, J. (1973), 'The Urban Field as a Human Habitat', in S. P. Snow (ed.), *The Place of Planning*, Auburn, Alabama, Auburn University Press, 32–44.
FSPA (Farm Shop and Pick Your Own Association) (1982), *Farm Fresh '82: The Guide to Farm Shops and Pick Your Own Farms throughout Britain*, London, FSPA.
Fuguitt, G.V., Voss, P.R. and Doherty, J.C. (1979), *Growth and Change in Rural America*, Washington, Urban Land Institute.
Fung, T. and Zhang, Q. (1989), 'Land Use Change Detection and Identification with Landsat Digital Data in the Kitchener–Waterloo Area', in C.R. Bryant, E.F. DeDrew, C. Marois and F. Cavayas (eds), *Remote Sensing and Methodologies of Land Use Change Analysis*, Waterloo, Canada, Department of Geography, University of Waterloo, Occasional Paper 6, 135–54.
Furuseth, O.J. (1980), 'The Oregon Agricultural Protection Program: A Review and Assessment', *Natural Resources Journal*, 20: 603–14.
Furuseth, O.J. (1981), 'Update on Oregon's Agricultural Protection Program: A Land Use Perspective', *Natural Resources Journal*, 21(1): 57–70.
Furuseth, O.J. (1985a), 'Influences on County Farmland Protection Efforts in California: A Discriminant Analysis', *Professional Geographer*, 37(4): 443–51.
Furuseth, O.J. (1985b), 'Local Farmland Conservation Programmes in the US: A Study of California Counties', *Applied Geography*, 5(3): 211–28.
Furuseth, O.J. and Pierce, J.T. (1982a), *Agricultural Land in an Urban Society*, Washington, Association of American Geographers.

BIBLIOGRAPHY

Furuseth, O.J., and Pierce, J.T. (1982b), 'A Comparative Analysis of Farmland Preservation Programmes in North America', *The Canadian Geographer*, 26(3): 191–206.

Gardner, B.D. (1977), 'The Economics of Agricultural Land Preservation', *American Journal of Agricultural Economics*, 58(12): 1027–36.

Gasson, R.M. (1968), 'Occupations Chosen by the Sons of Farmers', *Journal of Agricultural Economics*, 19: 317–26.

Gasson, R.M. (1973), 'Goals and Values of Farmers', *Journal of Agricultural Economics*, 24: 521–42.

Gayler, H.J. (1979), 'Political Attitudes and Urban Expansion in the Niagara Region', *Contact, Journal of Urban and Regional Affairs*, 11: 43–60.

Gayler, H.J. (1982), 'The Problems of Adjusting to Slow Growth in the Niagara Region of Ontario', *The Canadian Geographer*, 26(2): 165–72.

Gayler, H.J. (1991), 'The Role of Local Government in Urbanization and the Attrition of Agricultural Land', Miami, Fa., paper presented at the annual meeting of the Association of American Geographers, April 1991.

Geay, Y. (1974), 'L'utilisation des terres agricoles en matière d'urbanisation', Paris, Université Paris I, unpublished PhD thesis.

Gibson, J.A. (1977), 'On the Allocation of Prime Agricultural Land', *Journal of Soil and Water Conservation*, 32, (3): 271–5.

Gierman, D.M. (1977), *Rural to Urban Land Conversion*, Ottawa: Lands Directorate, Fisheries and Environment Canada, Occasional Paper 16.

Gilg, A.W. (1990), 'Farm Level Responses to Changes in the Common Agricultural Policy: A Report on Three Studies in South West England', Toronto, paper presented to the annual conference of the Association of American Geographers, April.

Gittinger, J., Leslie, J. and Hoisington, C. (eds) (1987), *Food Policy: Integrating Supply, Distribution and Consumption*, Baltimore, Johns Hopkins University Press.

Glenn, N.D. and Hill, L., Jr (1977), 'Rural–Urban Differences in Attitudes and Behaviour in the United States', *Annals of the American Academy of Political and Social Sciences*, 29(1): 36–50.

Gloudemans, R.J. (1971), *Use Value Farmland Assessment: Theory, Practice and Impact*, Chicago, International Association of Assessing Officials.

Goldman, G. and Strong, D. (1982), *Agricultural Land Use Control: A Short Analysis of Zoning, Taxing and Land Acquisition Techniques*, Sonoma County, Ca., University of California Cooperative Extension Service and Sonoma Farm Bureau.

Goodchild, R.N. and Munton, R.J.C. (1985), *Development and the Landowner: An Analysis of the British Experience*, London, George Allen and Unwin.

Graber, E.E. (1974), 'Newcomers and Oldtimers: Growth and Change in a Mountain Town', *Rural Sociology*, 39(4): 504–13.

Gramm, W.P. and Ekelund, R.B. (1975), 'Land Use Planning: The Market Alternative', in *No Man is an Island*, San Francisco, Institute for Contemporary Studies, 127–40.

Grasley, R.H. (1987), 'Understanding Entrepreneurs', *The Entrepreneurship Development Review*, 2: 24–5.

Greaves, S.M. (1984), 'Farmland Rental and Farm Enlargement: A Southern Ontario Example', Waterloo, Department of Geography, University of Waterloo, unpublished PhD thesis.

Gregor, H.F. (1963), 'Industrialised Drylot Farming: An Overview', *Economic Geography*, 39: 299–318.

Gregor, H.F. (1981), 'Agricultural Capitalization as a Spatial Complex', in N. Mohammad (ed.), *Perspectives in Agricultural Geography* (Volume 2), New Delhi, Concept Publishing Company, 407–26.

Gregor, H.F. (1982), *Industrialization of U.S. Agriculture: An Interpretative Atlas*, Boulder, Colorado, Westview Press.

Gregor, H.F. (1988), 'Urbanization and Agricultural Industrialization in Southern California', Auckland, New Zealand, paper presented to the IGU Commission on Changing Rural Systems, August.

Gregor, H.F. (1991), 'Metropolitan Agriculture as Industrializing Agriculture: The Orange County, CA, Example', Miami, Fa., paper presented at the annual meeting of the Association of American Geographers, April.

Grigg, D. (1984), *An Introduction to Agricultural Geography*, London, Hutchinson.

Grossman, M.G. (1987), 'Management Agreements in Dutch Agricultural Law: The

Contractual Integration of Agriculture and Conservation', *Denver Journal of International Law and Policy*, 16(1): 95–138.

Grossman, M.R. and Brussaard, W. (1987), 'The Land Shuffle: Reallocation of Agricultural Land under the Land Development Law in the Netherlands', *California Western International Journal*, 18(2): 209–89.

Grossman, M.R., and Brussaard, W. (1988), 'Planning, Development, and Management: Three Steps in the Legal Protection of Dutch Agricultural Land', *Washburn Law Journal*, 28(1): 86–149.

Grove-Hills, J., Munton, R.J.C. and Murdoch, J. (1990), *The Rural Land Development Process: Evolving a Methodology*, Countryside Change Working Paper Series 8, London, Department of Geography, University College.

Gustafson, G.C. (1977), *California's Use-Value Assessment Program: Participation and Performance through 1975–76*, Corvallis, Oregon, Economic Research Service, US Department of Agriculture.

Gustafson, G.C. and Wallace, L.T. (1975), 'Differential Assessment as Land Use Policy: The California Case', *Journal of the American Institute of Planners*, 41(6): 379–89.

Gwartney, J.D. (1976), *Economics: Private and Public Choice*, New York, Academic Press.

Hady, T.F. and Sibold, A.G. (1974), *State Programs for Differential Assessment of Farm and Open Space Land*, Washington, Economic Research Service, US Department of Agriculture, Agricultural Economics Research Report 256.

Hady, T.F. and Stinsin, T.F. (1967), *Taxation of Farmland on the Rural-Urban Fringe: A Summary of State Preferential Assessment*, Washington, Economic Research service, US Department of Agriculture, Agriculture Economics Research Report 119.

Hall, P.G. (1966), *Von Thunen's Isolated State*, Oxford, Pergamon Press.

Hall, P.G. (1988), *Cities of Tomorrow: An Intellectual History of Urban Planning and Design in the Twentieth Century*, Oxford, UK, Blackwell.

Hall, P., Gracey, H., Drewett, R. and Thomas R. (1973), *The Containment of Urban England: Volume Two: The Planning System*, London, George Allen and Unwin.

Hamill, A.E. (1969), 'Variables Related to Farm Real Estate Values in Minnesota Counties', *Agricultural Economics Research*, 21(1): 45–50.

Hansen, D.E. and Schwartz, S.I. (1975), 'Landowner Behaviour at the Rural–Urban Fringe in Response to Preferential Property Taxation', *Land Economics*, 51: 341–54.

Hansen, J.A.G. (1982), *Land Use Structure and Change in North America and the EEC*, Department of Environmental Studies and Countryside Planning, Wye College, University of London, Occasional Paper No. 6.

Hart, J.F. (1968), 'Loss and Abandonment of Cleared Farmland in the Eastern United States', *Annals of the Association of American Geographers*, 58(3): 417–40.

Hart, J.F. (1976), 'Urban Enchroachment on Rural Areas', *Geographical Review*, 66(1): 3–17.

Harvey, D.W. (1969), *Explanation in Geography*, London, Arnold.

Harvey, D.W. (1973), *Social Justice and the City*, London, Edward Arnold.

Held, R.B. and Visser, D.W. (1984), *Rural Land Uses and Planning: A Comparative Study of the Netherlands and the United States*, Amsterdam, Elsevier.

Hennigh, L. (1978), 'The Good Life and the Taxpayers' Revolt', *Rural Sociology*, 43(2): 178–89.

Hill, B. (1974), 'The Rise of the Mixed Tenure Farm: An Examination of Official Statistics', *Journal of Agricultural Economics*, 25(2): 177–82.

Hill, R.D. (1986), 'Land Use Change on the Urban Fringe', *Planning Quarterly*, 81: 15–6.

Hind-Smith, J. and Gertler, L.O. (1963), 'The Impact of Urban Growth on Agricultural Land: A Pilot Study', in *Resources for Tomorrow* (Background Papers: Supplementary Volume), Ottawa, The Queen's Printer, 155–80.

Hodge, G. (1974), 'The City in the Periphery', in L. S. Bourne, R. D. MacKinnon, J. Seigel and J. W. Simmons (eds), *Urban Futures for Central Canada: Perspectives on Forecasting Urban Growth and Change*, Toronto, University of Toronto Press, 281–301.

Hodge, G. (1987), 'Planning Rural Canada: Issues and Approaches', in P.J. Cloke (ed.), *Policies and Plans for Rural People: An International Perspective*, London, Unwin Hyman.

Hodge, G. and Qadeer, M.A. (1983), *Towns and Villages in Canada*, Toronto, Butterworth.

Hodge, I. and Whitby, M. (1981), *Rural Employment: Trends, Options, Choices*, London, Methuen.

Hoffman, D.W. (1971), *The Assessment of Soil Productivity for Agriculture*, Toronto, Ontario, Ministry of Agriculture and Food, ARDA Report 4.
Houée, P. (1990), 'Espaces ruraux: entre la décomposition et le renouvellement', *Aménagement Foncier Agricole et Rural*, 64(4): 18–26.
Huddleston, H. and Pease, J.R. (1979), *Criteria for Evaluating Tillamook County's Agricultural Lands*, Washington, DC, Soil Conservation Service, USDA.
Hushak, L.J. and Bovard, D.N. (1975), *The Demand for Land in the Urban-Rural Fringe*, Wooster, Ohio Agricultural and Development Centre, Research Bulletin 1076.
IAURIF (Institut d'Aménagement et d'Urbanisme de la Région d'Ile-de-France) (1878), *Etude sur le devenir des terres acquises par la SAFER à Dampierre*, Maincourt, Senlisse (78), Paris, IAURIF.
IAURIF (1987), *L'évolution des zones rurales en Ile-de-France de 1974 à 1982*, Paris, IAURIF.
Ilbery, B.W. (1983), 'Goals and Values of Hops Farmers', *Transactions of the Institute of British Geographers*, 8: 329–41.
Ilbery, B.W. (1985), *Agricultural Geography*, Oxford, Oxford University Press.
Ilbery, B.W. (1988a), 'Agricultural Change on the West Midlands Urban Fringe', *Tijdschrift voor Econ. en Soc. Geograpfie*, 79(2): 108–21.
Ilbery, B.W. (1988b), 'Farm Diversification and the Restructuring of Agriculture', *Outlook on Agriculture*, 17(1): 35–9.
Ilbery, B.W. (1990), 'Supply Control Measures and Farm Diversification in Britain', Toronto, paper presented to the annual conference of the Association of American Geographers, April.
INRS (Institut National de la Recherche Scientifique) (1973), *Région sud: l'agriculture*, Montreal, Office de Planification et du Développement du Québec and Université du Québec.
Ironside, R.G. (1979), 'Land Tenure, Farm Income and Farm Practice in Southern Ontario, Canada', *Ontario Geography*, 14: 21–39.
Irving, R.M. (1966), *Amenity Agriculture*, Vancouver, Tantalus Research Limited, BC Geographical Series 11.
Jackson, J.N. (1982), 'The Niagara Fruit Belt: The Ontario Municipal Board Decision of 1981', *The Canadian Geographer*, 26(2): 172–6.
Jackson, J.N. (1985), 'Protection of Land for Urban Use: A Reversal of the Canadian Norm in New Zealand', *The Canadian Geographer*, 29(4): 355–6.
Janick, J. (1979), *Horticultural Science*, San Francisco, W.H. Freeman.
Jenkins, W.I. (1978), *Policy Analysis*, New York, St Martin's Press.
Jesson, B. (1987), *Behind the Mirror Glass: The Growth of Wealth and Power in New Zealand in the Eighties*, Auckland, Penguin Books.
Johnson, D.D. and Howarth, P.J. (1989), 'The Effects of Spatial Resolution on Land Cover/Land Use Theme Extraction from Airborne Digital Data', in C.R. Bryant, E.F. LeDrew, C. Marois and F. Cavayas (eds), *Remote Sensing and Methodologies of Land Use Change Analysis*, Waterloo, Ontario, Department of Geography, University of Waterloo, Occasional Paper 6, 117–34.
Johnson, T.G., Marshall, J.P. and O'Dell, C.R. (1987), 'A Proposed Urban Agricultural Enterprise', in W. Lockeretz (ed.), *Sustaining Agriculture near Cities*, Ankeny, Iowa, Soil and Water Conservation Society, 37–47.
Johnston, T.R.R. (1983), 'An Evaluation of the Rationale of Farmland Preservation Policy in Ontario, Guelph, Ontario', Department of Geography, University of Guelph, unpublished MA thesis.
Johnston, T.R.R. (1989), 'Farmers' Adaptive Behaviour in an Urbanising Environment: Guelph to Toronto Area', Waterloo, Ontario, Department of Geography, University of Waterloo, unpublished PhD thesis.
Johnston, T.R.R. (1990), 'Subsumption and the Family Farm in New Zealand', Toronto, paper presented to the annual conference of the Association of American Geographers, April.
Johnston, T.R.R. and Bryant, C.R. (1987), 'Agricultural Adaptation: The Prospects for Sustaining Agriculture near Cities', in W. Lockeretz (ed.), *Sustaining Agriculture near Cities*, Ankeny, Iowa, Soil and Water Conservation Society, 9–21.
Johnston, T.R.R. and Bryant, C.R. (1989), 'Problems and Prospects of Farming in a Near-Urban Region: A Canadian Example', in R. Welch (ed.), *Proceedings of the Fifteenth New Zealand Geography Conference*, Dunedin, New Zealand Geographical Society 56–63.

Johnston, T.R.R. and Smit, B.E. (1985), 'An Evaluation of the Rationale for Farmland Preservation Policy in Ontario', *Land Use Policy*, 2, (3): 225–37.

Jones, G.E. (1963), 'The Diffusion of Agricultural Innovations', *Journal of Agricultural Economics*, 15(1): 49–59.

Jones, G.E. (1967), 'The Adoption and Diffusion of Agricultural Practices', *World Agricultural Economics and Rural Sociology Abstracts*, 9(1): 1–34.

Jones, R.C. (1976), 'Testing Macro-Thunen Models by Linear Programming', *The Professional Geographer*, 28(3): 353–61.

Joseph, A.E., Keddie, P.D. and Smit, B. (1988), 'Unravelling the Population Turnaround in Rural Canada', *The Canadian Geographer*, 32(1): 17–30.

Joseph, A.E. and Smit, B.E. (1981), 'Implications of Exurban Residential Development: A Review', *The Canadian Journal of Regional Science*, 4(2): 207–24.

Jumper, S.R. (1974), 'Wholesale Marketing of Fresh Vegetables', *Annals of the Association of American Geographers*, 64(2): 378–98.

Keating, M. (1986), 'Our Precious Fruitland in Peril: We're Covering them with Houses, Factories and Asphalt', *Canadian Geographic*, 106(5): 26–35.

Keng, C.B. (1976), 'Economic Implications Associated with Part-Time Farming', Palmerston North, New Zealand, Department of Agricultural Economics and Farm Management, Massey University, unpublished MAgrSc. thesis.

Kerstens, A.P.C. (1989), 'Land Development and Quality', Amsterdam, paper presented to the IGU Commission on Changing Rural Systems, Limits to Rural Land Use, August.

Kivlin, J.E. and Fliegel, F.C. (1968), 'Orientation to Agriculture: A Factor Analysis of Farmers' Perceptions of New Practices', *Rural Sociology*, 33: 127–40.

Klingebiel, A.A. and Montgomery, P.H. (1961), *Land Capability Classification*, Washington, Soil Conservation Service, USDA, Agricultural Handbook 210.

Kohn, W. (1990), 'Industrialised Agriculture and Land Use Conflicts in the Duemmer Lake Region', Toronto, paper presented to the annual conference of the Association of American Geographers, April.

Krueger, R.R. (1957), 'The Rural–Urban Fringe Taxation Problem: A Case Study of Louth Township', *Land Economics*, 33: 264–9.

Krueger, R.R. (1959), 'Changing Land Use Patterns in the Niagara Fruit Belt', *Transactions of the Royal Canadian Institute*, 32 (Part 2, No. 67): 39–140.

Krueger, R.R. (1968), 'The Geography of the Orchard Industry in Canada', in R.M. Irving (ed.), *Readings in Canadian Geography*, Toronto, Holt, Rinehart and Winston, 215–38.

Krueger, R.R. (1977a), 'The Destruction of a Unique Renewable Resource: The Case of the Niagara Fruit Belt', in R.R. Krueger and B. Mitchell (eds), *Managing Canada's Renewable Resources*, Toronto, Methuen, 132–48.

Krueger, R.R. (1977b), 'The Preservation of Agricultural Land in Canada', in R.R. Kruger and B. Mitchell (eds), *Managing Canada's Renewable Resources*, Toronto, Methuen, 119–31.

Krueger, R.R. (1978), 'Urbanisation of the Niagara Fruit Belt', *The Canadian Geographer*, 22: 179–94.

Krueger, R.R. (1982), 'The Struggle to Preserve Specialty Cropland in the Rural–Urban Fringe of the Niagara Peninsula of Ontario', *Environments*, 14, (3): 1–10.

Krueger, R.R., and Maguire, N.G. (1984), 'Changing Urban and Fruitgrowing Patterns in the Okanagon Valley, B.C.', *Environments*, 16: 1–9.

Land Use Advisory Council (1983), *Land Use in New Zealand: A National Goal*, Wellington, Department of Lands and Survey.

Lang, R., Boothroyd, P. and Armour A. (1976), *Impact and Controls Study: New CP/CN Railline and Yards for Regina*, Regina, City of Regina.

Lapping, M.B. and FitzSimons, J.F. (1982), 'Beyond the Land Issue: Farm Viability Strategies', *GeoJournal*, 6(6): 519–24.

Lapping, M.B., Penfold, G. and MacPherson, S. (1983), 'The Right to Farm Laws: Will They Resolve Land Conflicts?', *Journal of Soil and Water Conservation*, 38(6): 465–7.

Laureau, X. (1983), 'Agriculture péri-urbaine: des entreprises pour demain', *L'Agriculture d'Entreprise*, 171–72: 3–42.

Laureau, X. (1984), 'Agriculteurs péri-urbains: une nouvelle donne, de nouvelles opportunités', Beauvais, France, Institut Supérieur Agricole de Beauvais, mémoire de fin d'études.

Lawrence, H.W. (1988), 'Changes in Agricultural Production in Metropolitan Areas', *Professional Geographer*, 40(2): 159–75.
Layton, R.L. (1976), 'Hobby Farming: A Case Study of the Rural Urban Fringe of London', Ontario, London, Department of Geography, University of Western Ontario, unpublished MA thesis.
Layton, R.L. (1978), 'The Hobby Farm Issue', *Town and Country Planning*, 49: 53–4.
Leamy, M.L. (1974), 'Resources of Highly Productive Land', *New Zealand Agricultural Science*, 8: 187–91.
Lichfield, N., Kettle, P. and Whitebread, M. (1975), *Evaluation in the Planning Process*, Oxford, Oxford University Press.
Lin, S. and Labrosse, G. (1980), 'Canada's Agricultural and Food Trade in the 1970s', *Canadian Farm Economics*, 15(4): 1–8.
Lindblom, C.E. (1959), 'The Science of Muddling Through', *Public Administration Review*, 19(1): 79–88.
Lindblom, C.E. (1965), *The Intelligence of Democracy*, New York, Free Press.
Lindblom, C.E. (1973), 'The Science of Muddling Through', in A. Faludi (ed.), *A Reader in Planning Theory*, Oxford, Pergamon Press, 151–69.
Lipsey, R.G. (1983), *An Introduction to Positive Economics* (6th edn), London, Weidenfeld and Nicolson.
Listokin, D. (ed.) (1974) *Land Use Controls: Present Problems and Future Reform*, New Jersey, Center for Urban Policy Research, Rutgers University and State University of New Jersey.
Little, C.E. (1974), *The New Oregon Trail*, Washington, The Conservation Foundation.
Lockeretz, W. (1981), 'Crop Residues for Energy: Comparative Costs and Benefits for the Farmer, the Energy Facility, and the Public', *Energy in Agriculture*, 1: 71–89.
Lockeretz, W. (ed.) (1987), *Sustaining Agriculture near Cities*, Ankeny, Iowa, Soil and Water Conservation Society.
London Countryside Change Centre (1989), *The Countryside in Question: A Research Strategy*, London, Department of Geography, University College, Countryside Change Working Paper Series 1.
London and South East Regional Planning Conference (1987), *Developing SE Regional Strategic Guidance: Report of the Rural Issues Group*, London, The London and South East Regional Planning Conference.
Lowe, P., Marsden, T.K. and Whatmore, S.J. (1990), *Technological Change and the Rural Environment*, London, David Fulton Publishers, Critical Perspectives on Rural Change Series I.
Lowry, G.K., Jr (1980), 'Evaluating State Land Use Control: Perspectives and Hawaii Case Study', *Urban Land Annual*, 18: 85–127.
Lucas, P. and van Oort, G. (1991), 'Response of Farmers to the Loss of Land Caused by Urban Pressure', in G.M.R.A. van Oort, L.M. van den Berg, J.G. Groenendijk and A.H.H.M. Kempers (eds), *Limits to Rural Land Use*, Wageningen, The Netherlands, Centre for Agricultural Publishing and Documentation (Pudoc), 96–104.
Luzar, E.J. (1988), 'Strategies for Retaining Land in Agriculture: An Analysis of Virginia's Agricultural District Policy', *Landscape and Urban Planning*, 16(4): 319–32.
Macpherson, H. (1979), 'A Farmer/Rancher View of of Agricultural Land Retention Issues', in M. Schnepf (ed.), *Farmland, Food and the Future*, Ankeny, Iowa, Soil Conservation Society of America, 128–32.
Mage, J. (1982), 'The Geography of Part-Time Farming', *GeoJournal*, 6, (4): 301–12.
Mainié, P. and de Maillard, H. (1983), 'L'agriculture péri-urbaine: lieu privilégié d'expérimentation sociale', *Economie Rurale*, 155: 38–40.
Malassis, M. (1958), 'Economie des exploitations agricoles: essai sur les structures et les résultats des exploitations agricoles de grande et de petite superficie', Paris, Ecole Pratique des Hautes Etudes.
Mandelker, D. (1962). *Green Belts and Urban Growth*, Madison, Wisc., University of Wisconsin Press.
Manning, E.W. (1983), *Agricultural Land Protection Mechanisms in Canada*, Edmonton, Environmental Council of Alberta.
Manning, E.W. and Eddy, S.S. (1978), *The Agricultural Land Reserves of British Columbia: An*

Impact Analysis, Ottawa, Ontario, Lands Directorate, Environment Canada, Land Use in Canada Series 13.

Manning, E.W. and McCuaig, J.D. (1977), *Agricultural Land and Urban Centres*, Ottawa, Ontario, Lands Directorate, Environment Canada, Report 11.

Markusse, J.D. (1991), 'Possible Land-Use Changes at the Regional Level, Three Contrasting Scenarios for the Province of Friesland', in G.M.R.A. van Oort, L.M. van den Berg, J.G. Groenendijk and A.H.H.M. Kempers (eds), *Limits to Rural Land Use*, Wageningen, The Netherlands, Centre for Agricultural Publishing and Documentation (Pudoc), 63–76.

Marois, C., Bryant, C.R., Laurendeau, S. and Deslauriers, P. (1991), 'A Comparison of Regional Patterns of Agricultural Structure and Change: The Toronto and Montréal Fringes', Kingston, paper presented at the annual meeting of the Canadian Association of Geographers, June.

Marsden, T.K. et al. (Whatmore, S.J., Munton, R.J.C. and Little, J.K.) (1986a), 'The Restructuring Process and Economic Centrality in Capitalist Agriculture', *Journal of Rural Studies*, 2(4): 271–80.

Marsden, T.K. et al. (Munton, R.J.C., Whatmore, S.J. and Little, J.K.) (1986b), 'Towards a Political Economy of Capitalist Agriculture: A British Perspective', *International Journal of Urban and Regional Research*, 10(4): 498–521.

Marsden, T.K., Munton, R.J.C., Whatmore, S.J. and Little, J.K. (1989), 'Strategies for Coping in Capitalist Agriculture: An Examination of the Responses of Farm Families in British Agriculture', *Geoforum*, 20(1): 1–14.

Marsden, T.K., Lowe, P. and Whatmore, S.J. (1990), *Rural Restructuring: Global Processes and their Responses*, London, David Fulton Publishers, Critical Perspectives on Rural Change Series II.

Mason, R.J. (1991), 'Land Trusts as Shapers of Rural Landscapes', Miama, Fa., paper presented at the annual meetings of the Association of American Geographers, April.

Mattingly, D.F. (1972), 'Intensity of Agricultural Land Use near Cities', *The Professional Geographer*, 24(1): 7–10.

McAllister, D.M. (1982), *Evaluation in Environmental Planning: Assessing Environmental, Social, Economic and Political Trade-Offs*, Cambridge, The MIT Press.

McCallum, J. (1980), *Agriculture and Economic Development in Quebec and Ontario until 1870*, Toronto, University of Toronto Press.

McCuaig, J.D., and Manning, E.W. (1982), *The Evolution of Agricultural Land Use in Canada: Process and Consequences*, Ottawa, Ontario, Lands Directorate, Environment Canada, Land Use in Canada Series 21.

McDonald, G.T. (1974), 'The Fellowship of the Ring', in W. Brockie, Lettern, R. and Stokes, E. (eds), *Proceedings of the IGU Regional Conference*, Palmerston North, New Zealand Geographical Society, 33–42.

McDonald, G.T. (1989), 'Rural Land-Use Planning in Australia', in P.J. Cloke (ed.), *Rural Land-Use Planning in Developed Nations*, London, Unwin Hyman, 207–37.

McRae, J.D. (1977), 'Recent Changes in Land Ownership and Implications for Land Use: A Case Study in Eastern Ontario', Guelph, Department of Geography, University of Guelph, unpublished MA thesis.

McRae, J.D. (1980), *The Influence of Exurbanite Settlement on Rural Areas: A Review of the Canadian Literature*, Ottawa, Ontario, Lands Directorate, Environment Canada, Working Paper 3.

McRae, J.D. (1981), *The Impact of Exurban Settlement in Rural Areas: A Case Study in the Ottawa–Montreal Axis*, Ottawa, Ontario, Lands Directorate, Environment Canada, Working Paper 22.

Meister, A.D. (1981), 'Subdivision – The Rural County Councillor's Headache', *Town Planning Quarterly*, 63: 5–7.

Meister, A.D. (1982), *The Preservation and Use of Agricultural Land: Land Use Policies and Their Implementation*, Palmerston North, New Zealand, Department of Agricultural Economics and Farm Management, Massey University, Discussion Paper in Natural Resource Economics 5.

Meister, A.D. and Stewart, D.S. (1980), *A Study of Rural Small Holdings in Taranaki County*, Palmerston North, New Zealand, Department of Agricultural Economics and Farm Management, Massey University, Discussion Paper in Natural Resource Economics 3.

Melbourne and Metropolitan Board of Works (1977), 'A Multiplicity of Views', *Living Together*, 21: 9–11.
Merget, A.E. (1981), 'Achieving Equity in an Era of Fiscal Restraint', in R.W. Burchell and D. Listokin (eds), *Cities under Stress: The Fiscal Crises of Urban America*, Rutgers, The Center for Urban Policy Research.
Meyerson, M. and Banfield, E.C. (1955), *Politics, Planning and the Public Interest*, Glencoe, Free Press.
Mitchell, G.F.C. (1969), *Application of a Likert-Type Scale to the Measurement of the Degree of Farmers' Subscription to Certain Goals or Values*, Department of Economics, University of Bristol.
Mitchell, B. (1979), *Geography and Resource Analysis*, London, Longman.
Mitchell, D. (1975), *The Politics of Food*, Toronto, James Lorimer.
Molloy, L.F. (1980), *Land Alone Endures: Land Use and the Role of Research*, Wellington, New Zealand, New Zealand Department of Scientific and Industrial Research.
Molnar, J.J. (1985), 'Farmland Protection as a Community Development Issue: Alabama Farmer and Landowner Perspectives', *Journal of Soil and Water Conservation*, 40(6): 528–31.
Moncriff, P.M. and Phillips, W.E. (1972), 'Rural Urban Interface Acreage Developments: Some Observations and Policy Implications', *Canadian Journal of Agricultural Economics*, 20(1): 80–4.
Mooney, P.F. (1990), 'Re-examining Preservation of Agricultural Land in B.C.', in R. Corbett (ed.), *Protecting our Common Future: Conflict Resolution within the Farming Community*, Sackville, New Brunswick, Rural and Small Town Research and Studies Programme, Mount Allison University, 1–34.
Moore, A. (1990), 'Viable Agricultural Alternatives: Challenges and Opportunities for Family Farm Units', *AgriScience*, July/August: 4–5.
Moore, C.V. and Dean, G.W. (1972), 'Industrialized Farming', in A.G. Ball and E.O. Heady (eds), *Size, Structure and Future of Farms*, Ames, Iowa State University Press, 214–31.
Moran, W. (1978), 'Land Value, Distance and Productivity on the Auckland Urban Periphery', *New Zealand Geographer*, 34(2): 85–96.
Moran, W. (1979), 'Spatial Patterns of Agriculture on the Urban Periphery: The Auckland Case', *Tijdschrift voor Econ. en Soc. Geografie*, 70(3): 164–76.
Moran, W. (1980), 'Mechanisms of Land Use Allocation on the Urban Periphery', in G. Anderson (ed.), *The Land Our Future*, Auckland, Longman, 223–38.
Moran, W. (1987), 'Marketing Structures and Rural Land Use Change', *New Zealand Geographer*, December: 164–9.
Moran, W. (1988), 'The Farm Equity Cycle and Enterprise Choice', *Geographical Analysis*, 20: 84–91.
Moran, W., Benediktsson, K. and Manning, S. (1989), 'Labour Processes and Family Farms', Dunedin, New Zealand, paper presented at the 15th Conference of the New Zealand Geographical Society, Otago University.
Moran, W. and Nason, S.J. (1982), 'Spatio-Temporal Localization of New Zealand Dairying', *Australian Geographical Studies*, 19(1): 47–66.
Moran, W., Neville, W. and Rankin, D. (1980), *Agriculture and Productivity of Small Holdings*, Auckland, New Zealand, Auckland Regional Authority.
Morgan, W.B. and Munton, R.J.C. (1971), *Agricultural Geography*, London, Methuen.
Morris, A. (1989), 'The Restructuring of Apple Varieties Within the Apple Industry, Concentrating on the Hawkes Bay Region', Palmerston North, New Zealand, Department of Geography, Massey University, undergraduate research project.
Morris, E.M. (1981), 'The Preservation of the Rural and Cultural Landscape: Challenges and Techniques in Preserving Historic Places and Open Spaces', *Pennsylvania Geographer*, 19(2): 32–6.
Munton, R.J.C. (1974), 'Farming on the Urban Fringe', in J.H. Johnson (ed.), *Suburban Growth*, London, John Wiley, 201–63.
Munton, R.J.C. (1976), 'An Analysis of Price Trends in the Agricultural Land Market of England and Wales', *Tijdschrift voor Econ. en Soc. Geografie*, 67(4): 202–12.
Munton, R.J.C. (1983a), *London's Green Belt: Containment in Practice*, Chichester, UK, John Wiley.
Munton, R.J.C. (1983b), 'Agriculture and Conservation: What Room for Compromise?', in A. Warren and F.B. Goldsmith (eds), *Conservation in Perspective*, Chichester, UK, John Wiley, 353–73.

Munton, R.J.C. (1985), 'Investment in British Agriculture by the Financial Institutions', *Sociologia Ruralis*, 25: 153–73.

Munton, R.J.C. (1987), 'The Conflict Between Conservation and Food Production in Great Britain', in C. Cocklin, B.E. Smit and T.R.R. Johnston (eds), *Demands on Rural Lands: Planning for Resources Use*, Boulder, Westview, 47–60.

Munton, R.J.C., Whatmore, S., and Marsden, T. (1988), 'Reconsidering Urban-Fringe Agriculture: A Longitudinal Analysis of Capital Restructuring on Farms in the Metropolitan Green Belt', *Transactions of the Institute of British Geographers*, 13: 324–36.

Musgrave, R.A. (1959), *The Theory of Public Finance: A Study in Public Economy*, New York, McGraw-Hill.

Muth, R.F. (1961), 'Economic Change and Rural–Urban Land Use Conversions', *Econometrica*, 29(1):1–23.

Naisbitt, J. (1982), *Megatrends: Ten New Directions Transforming our Lives*, New York, Warner Books.

NALS (National Agricultural Lands Study) (1980), *The Protection of Farmland: Executive Summary*, Washington, US Government Printing Office.

NALS (National Agricultural Lands Study) (1981), *Final Report*, Washington, US Government Printing Office.

Napton, D. (1990), 'Regional Farmland Protection: The Twin Cities Experience', *Journal of Soil and Water Conservation*, July–August: 446–9.

National Capital Commission (1982), *The Management Plan for the Greenbelt*, Ottawa, National Commission Commission.

Neimanis, V.P. (1979), *Canada's Cities and Their Surrounding Land Resources*, Ottawa, Ontario, Lands Directorate, Environment Canada, Canada Land Inventory Report 15.

Nelson, A.C. (1990), 'An Appraisal of Farmland Preservation Policies', in R. Corbett (ed.), *Protecting our Common Future: Conflict Resolution within the Farming Community*, Sackville, New Brunswick, Rural and Small Town Research and Studies Programme, Mount Allison University, 79–109.

Nelson, R.H. (1977), *Zoning and Property Rights: An Analysis of the American System of Land Use Regulation*, Cambridge, Mass., MIT Press.

OECD (Organisation of Economic Co-operation and Development) (1976), *Agriculture in the Planning and Management of Peri-Urban Areas*, Paris, OECD.

Olmstead, C.W. (1970), 'The Phenomena, Functioning Units and Systems of Agriculture', *Geographia Polonica*, 19(1): 31–42

One Thousand Friends of Oregon (1980), *Four Year Report 1975-1979*, Portland, Oregon, 1000 Friends of Oregon.

Ontario Institute of Agrologists (1975), *Foodland Preservation or Starvation*, Hillsburgh, Ontario Institute of Agrologists.

Ontario Ministry of Agriculture and Food (1976), *Agricultural Code of Practice*, Toronto, Ontario Ministry of Agriculture and Food.

Ontario Ministry of Agriculture and Food (1984), *1984 Pick-Your-Own Fruit and Vegetable Farms and Farmers' Markets*, Toronto, Ontario Ministry of Agriculture and Food.

Orhon, J. (1982), 'L'espace péri-urbain: un nouvel espace?', *Etats Fonciers*, 14: 4pp.

O'Riordan, T. (1971), *Perspectives on Resource Management*, London, Pion.

O'Riordan, T. (1986), 'Halvergate: Anatomy of a Decision', in J.T. Coppock and P. Kivell (eds), *Geography: Planning and Policy Making?*, Norwich, Geo Books, 189–228.

Pacione, M. (ed.) (1984), *Rural Geography*, London, Harper and Row.

Pacione, M. (ed.) (1986), *Progress in Agricultural Geography*, Beckenham, UK, Croom Helm.

Pahl, R.E. (1975), *Whose City?*, London, Longman.

Pahl, R.E. (1977), 'Managers, Technical Experts and the State: Forms of Mediation, Manipulation and Dominance in Urban and Regional Development', in M. Harloe (ed.), *Captive Cities*, London, John Wiley.

Parikh, B. and Rabar, F. (1981), 'Food Problems and Policies: Present and Future, Local and Global', in K. Parikh and F. Rabar (eds), *Food for All in a Sustainable World: The IIASA Food and Agriculture Program*, Laxenburg, Austria, International Institute for Applied Systems Analysis, 1–23.

Pautard, J. (1965), *Les disparités régionales dans la croissance de l'agriculture française*, Paris, Gauthier-Villars, Série Espace Economique.

Pease, J.R. (1982), 'Commercial Farmland Preservation in Oregon', *GeoJournal*, 6(6): 547–53.
Pease, J.R. (1983), 'Regional Characteristics of Commercial Agriculture in Oregon', Corvallis, Oregon, Oregon State University, unpublished paper.
Pease, J.R. (1990), 'Land Use Designation in Rural Areas: An Oregon Case Study', *Journal of Soil and Water Conservation*, September–October: 524–8.
Pease, J.R. and Morgan, M. (1979), *Community Growth Management: Performance Zoning*, Corvallis, Oregon, Oregon State University Extension Service, Extension Circular 963.
Pédelaborde, P. (1961), *L'agriculture dans les plaines alluviales de la presqu'île de Saint-Germain-en-Laye*, Paris, A. Colin.
Penfold, G. (1990), 'Right-to-Farm as a Method of Conflict Resolution', in R. Corbett (ed.), *Protecting our Common Future: Conflict Resolution within the Farming Community*, Sackville, New Brunswick, Rural and Small Town Research and Studies Programme, Mount Allison University, 63–78.
Philippe, F. and Biancale, M. (1981), *La sécurité alimentaire en région Ile-de-France*, Paris, IAURIF.
Phillips, A. (1985), 'Conservation at the Crossroads: The Countryside', *Geographical Journal*, 151(2): 237–45.
Phipps, T. (1983), 'Landowner Incentives to Participate in a Purchase of Development Rights Program with Application to Maryland', *Journal of the Northeastern Agricultural Economic Council*, 12(1): 61–5.
Phlipponneau, M. (1956), *La vie rurale de la banlieue parisienne: étude de géographie humaine*, Paris, Ecole Pratique des Hautes Etudes.
Pierce, J.T. (1981a), 'The BC Agricultural Land Commission: Review and Evaluation', *Plan Canada*, 21(2): 48–56.
Pierce, J.T. (1981b), 'Conversion of Rural Land to Urban: A Canadian Profile', *Professional Geographer*, 21(2): 163–73.
Pierce, J.T. (1987), 'Risk-Aversion Versus Optimal Strategies for Planning Rural Resources', in C. Cocklin, B.E. Smit and T.R.R. Johnston (eds), *Demands on Rural Lands: Planning for Resource Use*, Boulder, Westview Press, 113–22.
Pierce, J.T. (1991), 'Agriculture, Sustainability and the Imperatives of Policy Reform', Miami, Fa., paper presented at the annual meetings of the American Association of Geographers, April.
Pigram, J.J. (1987), 'Countryside Parks and Multipurpose Use of Rural Resources', in C. Cocklin, B.E. Smit and T.R.R. Johnston (eds), *Demands on Rural Lands: Planning for Resource Use*, Boulder, Westview Press, 61–70.
Pinault, M. (1983), 'Allez aux fraises! dans les jardins de Cergy—Pontoise', *La France Agricole*, 58–9.
Pitt, D.G., Lessley, B.V. and Phipps, T. (1988), 'Influences of Local and Region Population Trends and Development Pressures in Maryland's Agricultural Land Preservation Program', *Landscape and Urban Planning*, 15(3–4): 337–49.
Platt, R.H. (1972), *The Open Space Decision Process*, Chicago, Department of Geography, University of Chicago, Research Paper 142.
Ploch, L.A. (1978), 'The Reversal in Migration Patterns: Some Rural Development Consequences', *Rural Sociology*, 43(2): 293–303.
Plumb, T. (1980), *Atlas of Australian Resources*, Third Series, Vol. 2: Population, Canberra, Division of National Mapping.
Plumb, T. (1982), *Atlas of Australian Resources*, Third Series, Vol. 3: Agriculture, Canberra, Division of National Mapping.
Pomeroy, A. (1986), 'A Sociological Analysis of Structural Change in Pastoral Farming in New Zealand', Colchester, UK, Department of Sociology, University of Essex, unpublished PhD thesis.
Pred, A.R. (1967), *Behaviour and Location: Foundations for a Geographic and Dynamic Location Theory*, Part 2, Stockholm, University of Lund, Lund Studies in Geography, Series B, 28.
Préfecture de la Région d'Ile-de-France (1976), *Schéma Directeur d'Aménagement et d'Urbanisme de la Région d'Ile-de-France*, Paris, Service Régional de l'Equipement de la Région d'Ile-de-France and Institut d'Aménagement et d'Urbanisme de la Région d'Ile-de-France.
Préfecture de la Région d'Ile-de-France (1988) *Les exploitations spécialisées en Ile-de-France*, Paris, Préfecture de la Région d'Ile-de-France.
Préfecture de la Région d'Ile-de-France (1991), *L'Ile-de-France au futur: esquisse du nouveau schéma*

directeur de l'Ile-de-France, Paris, Préfecture de la Région d'Ile-de-France, Direction régionale de l'équipement.

Preston, R.E., Dudycha, D.J. and Goldmann, G.J. (1987) The Waterloo Generic Urban Model (WATGUM) Project and Workshop, *Environments*, 19(1: Special Issue).

Price, L.W. (1981), *Mountains and Man: A Study of Process and Environment*, Berkeley, University of California Press.

Pryor, R.J. (1968), 'Defining the Rural–Urban Fringe', *Social Forces*, 47: 202–15.

Punter, J.V. (1976), *The Impact of Exurban Development on Land and Landscape in the Toronto-Centred Region, 1954–1971*, Ottawa, Central Mortgage and Housing Corporation, Policy Planning Division.

Putnam, R.G. (1962), 'Changes in Rural Land Use Patterns on the Central Lake Ontario Plain', *The Canadian Geographer*, 6(1): 60–8.

Pyle, L.A. (1989), 'Persistent Landsownership at the Rural–Urban Fringe', *Urban Geography*, 10(2): 157–71.

Rajotte, F. (1973), 'The Quebec City Recreational Hinterland', Montréal, Department of Geography, McGill University, unpublished PhD thesis.

Rancich, M.T. (1970), 'Land Value Change in an Area Undergoing Urbanization', *Land Economics*, 40(1): 32–40.

Ratcliffe, J. (1974), *An Introduction to Town and Country Planning*, London, Hutchinson.

Rawson, M. (1977), 'Letter to the Department of City Planning, Faculty of Agriculture, University of Manitoba', quoted in C. Beaubien and R. Tabacnik, *People and Agricultural Land*, Perceptions 4, Ottawa, Science Council of Canada.

Reeds, L.G. (1969), *Niagara Region: Agricultural Research Report*, Toronto, Ontario Ministry of Treasury, Economics and Intergovernmental Affairs.

Rees, R. (1984), *Public Enterprise Economics* (2nd edn), London, Weidenfeld and Nicolson.

Reynolds, L.G. (1966), *Economics: A General Introduction*, Homewood, Richard D. Irwin Inc.

Rich, R.C. (1979), 'Neglected Issues in the Study of Urban Service Distributions: A Research Agenda', *Urban Studies*, 16: 143–56.

Rickard, T.J. (1991a), 'Direct Marketing as Agricultural Adaptation in Megalopolitan Connecticut', in G.M.R.A. van Oort, L.M. van den Berg, J.G. Groenendijk and A.H.H.M. Kempers (eds), *Limits to Rural Land Use*, Wageningen, The Netherlands, Centre for Agricultural Publishing and Documentation (Pudoc), 78–88.

Rickard, T.J. (1991b), 'Public Policy and Agricultural Restructuring in Connecticut's Rural–Urban Fringe', Miami, Fa., paper presented at the annual meeting of the Association of American Geographers, April.

Rickson, R.E. and Neumann, R. (1984), *Farmers' Responses to Land Use Planning in Moreton and Boonah Shires*, Brisbane, Institute of Applied Social Research.

Roberts, N.A., and Brown, H.J. (1980), *Property Tax Preferences for Agricultural Land*, Monclair, New Jersey, Land Mark Studies, Lincoln Institute of Land Policy, Allanheld, Osmun and Co.

Robinson, D.A. (1989), 'Soil Erosion, Soil Conservation and Agricultural Policy for Arable Land in the U.K.', *Geoforum*, 20(1): 83–92.

Robinson, K. (1968), *The Law of Town and Country Planning*, Wellington, Butterworth.

Rodd, R.S. (1974), 'A Remarkable Change in the Rural Land Market', Guelph, University of Guelph, *Notes on Agriculture*, 12(2): 21.

Rodd, R.S. (1976), 'The Crisis of Land in the Ontario Countryside', *Plan Canada*, 10: 367–72.

Rose, J.G. (1984), 'Farmland Preservation Policy and Programs', *Natural Resources Journal*, 24(3): 591–640.

Russwurm, L.H. (1970), *Development of an Urban Corridor System, Toronto to Stratford Area, 1941–1966*, Toronto, The Queen's Printer, Regional Development Branch Research Paper 3.

Russwurm, L.H. (1977), *The Surroundings of Our Cities*, Ottawa, Community Planning Press.

Rutherford, J. (1966), 'Farming in the Sydney Region', in J. Rutherford et al. (eds), *New Viewpoints in Economic Geography*, Sydney, Martindale Press, 245–73.

Ruttan, V.W. (1955), 'The Impact of Urban-Industrial Development on Agriculture in the Tennessee Valley and the Southeast', *Journal of Farm Economics*, 37: 38–56.

Samuelson, P.A. (1954), 'The Pure Theory of Public Expenditure', *The Review of Economics and Statistics*, 36: 387–9.

Samuelson, P.A. (1955), 'Diagrammatic Exposition of a Theory of Public Expenditure', *The Review of Economics and Statistics*, 37: 350–6.

Sargent, C.A. (1970), *Urbanization of a Rural County*, Lafayette, Purdue University, Agricultural Experimental Station, Research Bulletin 859.
Schwarz, S.I., Hansen, D.E. and Foin, T.C. (1976), 'Landowner Benefits from Use-Value Assessment under the California Land Conservation Act', *American Journal of Agricultural Economics*, 58(2): 170–8.
Schultz, T.W. (1951), 'A Framework for Land Economics: The Long View', *Journal of Farm Economics*, 33: 204–15.
Schultz, T.W. (1953), *The Economic Organisation of Agriculture*, London, McGraw-Hill.
Senate Committee (1984), *Soil at Risk: Canada's Eroding Future*, Ottawa, Standing Committee of Senate on Agriculture, Fisheries and Forests.
Seni, D.A. (1978), 'Urban Plan and Policy Evaluation: An Assessment of Trends and Needs', *Plan Canada*, 18(2): 105–17.
Sewell, W.R.D. (1983), 'When do Environmental Issues Become Political Priorities?', seminar presented at the University of Guelph, Guelph, Ontario, March 1978.
Sewell, W.R.D. and Coppock, J.T. (1976), 'Achievements and Prospects', in J.T. Coppock and W.R.D. Sewell (eds), *Spatial Dimensions of Public Policy*, Oxford, Pergamon Press, 257–62.
Simpson-Lewis, W., Moore, J.E., Pocock, N.J., Taylor, M.C. and Swan, H. (1979), *Canada's Special Resource Lands: A National Perspective of Selected Land Uses*, Ottawa, Ontario, Lands Directorate, Environment Canada.
Sinclair, R.J. (1967), 'Von Thunen and Urban Sprawl', *Annals of the Association of American Geographers*, 57: 72–87.
Sisler, D.G. (1959), 'Regional Differences in the Impact of Urban–Industrial Development on Farm and Non-Farm Income', *Journal of Farm Economics*, 41: 1100–12.
Skoretz, P.W. (1990), 'An Evaluation of the Farm Tax Rebate Program's Role in Preserving Agricultural Land on the Rural–Urban Fringe', Waterloo, Ontario, School of Urban and Regional Planning, University of Waterloo, unpublished MA thesis.
Sly, W.K. (1970), 'A Climatic Moisture Index and Soil Classification', *Canadian Journal of Soil Sciences*, 50: 291–301.
Smit, B.E. (1979), 'Regional Employment Changes in Canadian Agriculture', *The Canadian Geographer*, 23(1): 1–17.
Smit, B.E. (1981), 'Prime Land, Land Evaluation, and Land Use Policy', *Journal of Soil and Water Conservation*, 36: 209–12.
Smit, B., Bond, D. and Brklacich, M. (1984), 'A Land Evaluation System: Design and Application in Canada', in J.W. Frazier (ed.), *Papers and Proceeding of Applied Geography Conferences*: Vol. 7, Binghampton, Department of Geography, State University of New York, 1–9.
Smit, B.E. and Cocklin, C. (1981), 'Future Urban Growth and Agricultual Land: Alternatives for Ontario', *Ontario Geography*, 18: 47–55.
Smit, B.E. and Flaherty, M.F. (1980), 'Preferences for Rural Land Severances: An Empirical Analysis', *The Canadian Geographer*, 24(2): 165–76.
Smit, B.E. and Flaherty, M.F. (1981), 'Resident Attitudes toward Exurban Development in a Rural Ontario Township', *The Professional Geographer*, 33(1): 103–12.
Smit, B.E. and Johnston, T.R.R. (1983), 'Public Policy Assessment: Evaluating Objectives of Resource Policies', *The Professional Geographer*, 35(2): 172–78.
Smit, B.E., Johnston, T.R.R. and Morse, R. (1985), 'Labour Turnover on Flue-Cured Tobacco Farms in Southern Ontario', *Agricultural Adminstration*, 20(3): 153–68.
Smit, B.E., Ludlow, L., Johnston, T.R.R. and Flaherty, M. (1987), 'Identifying Important Agricultural Lands: A Critique', *The Canadian Geographer*, 31(4): 356–65.
Smit, B.E., Rodd, S., Bond, D., Brklacich, M., Cocklin, C. and Dyer, A. (1983), 'Implications for Food Production Potential of Future Urban Expansion in Ontario', *Socio-Economic Planning Sciences*, 17(3): 109–19.
Smith, D.L. (1972), 'The Growth and Stagnation of an Urban Fringe Market Gardening Region – Virginia, South Australia', *Australian Geographer*, 12(1): 35–48.
Smith, S.N. (1987), 'Farming near Cities in a Bi-modal Agriculture', in W. Lockeretz (ed.), *Sustaining Agriculture near Cities*, Ankeny, Iowa, Soil and Water Conservation Society, 77–90.
Snyder, J.H. (1966), 'A New Programme for Agricultural Land Use Stabilisation: The California Land Conservation Act of 1965', *Land Economics*, 42: 29–41.

Spaulding, B. and Heady, E.O. (1977), 'Future Use of Agricultural Land for Nonagricultural Uses', *Journal of Soil and Water Conservation*, 32(1): 88–93.
Special Committee on Farm Income (Ontario) (1969), *Farm People in Ontario*, Toronto, The Queen's Printer, Special Committee on Farm Income, Research Report 5.
Spooner, J.W. (1966), 'Urban Development in Rural Areas', Sarnia, Ontario, address by the Minister of Housing to the Annual Meeting of the Association of Ontario Mayors and Reeves, 27 June.
Standing Conference (1976), The Improvement of London's Green Belt, London, Standing Conference on London and South-East Regional Planning.
Stockham, J. and Pease, J.R. (1974), *Performance Standards – A Technique for Controlling Land Use*, Corvallis, Oregon, Oregon State University Extension Service, Special Report 424.
Stonyer, E.J. (1973), 'The Conservation of Land for Primary Production', *New Zealand Surveyor*, 23: 133–40.
Storie, R.E. (1964), *Handbook on Soil Evaluation*, Berkeley, California, Agricultural Experimental Station, College of Agriculture, University of California.
Strachan, A. (1974), 'The Planning Framework for Modern Urban Growth: The Example of Great Britain', in J.H. Johnson (ed.), *Suburban Growth*, London, John Wiley, 53–76.
Sublett, M.D. (1975), *Farmers on the Road*, Chicago, Department of Geography, University of Chicago, Research Paper 168.
Sullivan, J.P. (1977), *Agricultural Districts: The New York Experience in Farmland Preservation*, in *Land Use: Tough Choices in Today's World*, Ankeny, Iowa, Soil Conservation Society of America.
Sutcliff, A. (1980), *The Rise of Modern Urban Planning, 1800–1914*, New York, St Martins Press.
Swain, P.J. and Haigh, V.J. (1985), *Farming and the Countryside*, London, UK Ministry of Agriculture, Fisheries and Food, Booklet 2384.
Taylor, C.C. (1949), 'Farm People's Attitudes and Opinions', in E.A. Schuller and C.C. Taylor (eds), *Rural Life in the United States*, New York, A. Knopf.
Thibodeau, J.C. (1984), 'Une urbanisation mieux contenue, une agriculture qui se régénère', *Cahiers de l'Institut d'Aménagement et d'Urbanisme de la Région d'Ile-de-France*, 73: 26–39.
Thibodeau, J.C., Gaudreau, M. and Bergeron, J. (1986), *Le zonage agricole, un bilan positif*, Montreal, Institut National de la Recherche Scientifique-Urbanisation, Research Report 9.
Thomas, D. (1970), *London's Green Belt*, London, Faber and Faber.
Thomson, K.J. (1981), *Farming in the Fringe: An Analysis of Agricultural Census Data Drawn from Parishes around the Six Metropolitan Counties and London*, Cheltenham, UK, Countryside Commission, CCP 142.
Toch, S.L. (1988), 'Resource Preservation through Community Integration: Towards an Enabling Environment (Waterloo and the Rhodon Valley, France)', Waterloo, Department of Geography, University of Waterloo, unpublished MA thesis.
Toner, W. (1979), 'Local Programs to Save Farms and Farmlands', in M. Schepf (ed.), *Farmland, Food and the Future*, Ankeny, Soil and Water Conservation Society of America, 189–202.
Tousaw, S.B. (1991), 'Conservation Tillage and the Rural Community', Waterloo, Ontario, Department of Geography, University of Waterloo, unpublished MA thesis.
Tricart, J. (1951), *La culture fruitière dans la région parisienne*, Paris, Centre National de la Recherche Scientifique, Etudes et Mémoires, Volume 2.
Troughton, M.J. (1976a), *Landholding in a Rural-Urban Fringe Environment: The Case of London*, Ontario, Ottawa, Environment Canada, Occasional Paper 11.
Troughton, M.J. (1976b), 'Comparative Profiles of Land Holding Types in the Rural–Urban Fringe of London', *Ontario Geography*, 10: 27–53.
Troughton, M.J. (1982a), 'Process and Response in the Industrialisation of Agriculture', in G. Enyedi and I. Volgyes (eds), *The Effects of Modern Agriculture on Rural Development*, New York, Pergamon Press, 213–27.
Troughton, M.J. (1982b), *Canadian Agriculture*, Budapest, Hungarian Academy of Sciences, Research Institute of Geography, Akademiai Kiado, Geography of World Agriculture 10.
Troughton, M.J. (1986), 'Rural Canada: What Future?', paper presented at the conference *Integrated Development Beyond the City*, Mount Allison University, Sackville, New Brunswick.
Trzyna, T.C. (ed.) (1984), *Preserving Agricultural Lands: An International Annotated Bibliography*, Berkeley, California Institute of Public Affairs, Environmental Studies Series 7.

Tweeten, L. (1983), 'The Economics of Small Farms', *Science*, 219(4588): 1037–41.
UK Ministry of Fisheries and Food (1977), 'Peri-Urban Agriculture in the Slough–Hillingdon Area (Region of London)', Paris, OECD, paper presented at the conference, *Peri-Urban Agriculture*.
Vachon, B. (1988), 'Quelques aspects géographiques financiers et politiques du zonage agricole au Québec', *Recherches Sociographiques*, 29(2–3): 417–30.
Vail, D. (1987), 'Suburbanization of the Countryside', in W. Lockeretz (ed.), *Sustaining Agriculture near Cities*, Ankeny, Iowa, Soil and Water Conservation Society, 23–36.
van den Berg, L.M. (1991), 'Quasi-Agricultural Land: Hidden Urbanization, Hobby Farming or What Else?' in G.M.R.A. van Oort, L.M. van den Berg, J.G. Groenendijk and A.H.H.M. Kempers (eds), *Limits to Rural Land Use*, Wageningen, The Netherlands, Centre for Agricultural Publishing and Documentation (Pudoc), 130–38.
van den Berg, L.M. and Ijkelenstam, G.F.P. (1983), *Land-Use Dynamics in the Rurban Fringe. Two Case Studies Compared: Lusaka (Zambia) and Haaglanden (Netherlands)*, Wageningen, The Netherlands, Institute for Land and Water Management Research Report 11.
van Oort, G. (1984a), 'L'aménagement rural dans la région nord d'Utrecht: le cas du Noorderpark', *Hommes et Terres du Nord*, 4: 273–80.
van Oort, G. (1984b), 'L'aménagement du territoire aux Pays-Bas: l'évolution de la Randstad', *Hommes et Terres du Nord*, 4: 226–36.
VBA (Verenigde Bloemenveilingen Aalsmeer) (1989), *Aalsmeer Flower Auction*, Amsterdam, Cooperative Association VBA.
Vining, D.R., Jr, Plaut, T. and Bieri, K. (1977), 'Urban Encroachment on Prime Agricultural Land in the United States', *International Regional Science Review*, 2:143–56.
Vogelar, I. (1978), 'The Effectiveness of Differential Assessment of Farmland in Metropolitan Chicago', *Geographical Survey*, 7:23–32.
Wagner, C. (1975), *Rural Retreats: An Urban Paper*, Canberra, Department of Urban and Regional Development – AGPS.
Walker, G. (1977), 'Social Networks and Territory in a Commuter Village: Bond Head, Ontario', *The Canadian Geographer*, 21(4): 329–50.
Walker, G. (1987), *The Invasion of the Countryside*, Toronto, York University/Atkinson College, Geographical Monograph 17.
Warren, L. and Rump, P. (1981), *Urbanization in Canada, 1966–1976*, Ottawa, Ontario, Lands Directorate, Environment Canada, Land Use in Canada Series 20.
WCED (World Commission on Environment and Development) (1987), *Our Common Future*, Oxford, Oxford University Press.
Weeds, Trees and Turf (1971), 'Canadian Sod Giant: Fairlawn is a Study in Management and Marketing', *Weeds, Trees and Turf*, November, reprint, 3pp.
Weeks, S.D. (1973), 'Rural Ruination or Rural Renaissance', *Small Town*, 6(1): 4–6.
Wellington (County) Planning and Development Department (1977), *An Analysis of the Financial Impact of Non-Farm Development upon Erin Township, 1966–1976*, Guelph, Wellington County Planning and Development Department.
Whatmore, S., Munton, R.J.C., Little, J. and Marsden, T. (1986), 'Internal and External Relations in the Transformation of the Family Farm', *Sociologia Ruralis*, 26(3/4): 396–8.
Whitby, M.C., Robins, D.L.J., Tansey, A.W. and Willis, K.G. (1974), *Rural Resource Development*, London, Methuen.
Wibberley, G.P. (1959), *Agriculture and Urban Growth: A Study of the Competition for Rural Land*, London, Michael Joseph.
Williams, D.B., and MacAulay, T.G. (1971), 'Changes in Rural Population and Work Force in Victoria, 1961–66', *Australian Geographical Studies*, 9: 161–71.
Williams, E.A. (1969), *Open Space: The Choices before California*, San Francisco, Diablo Press, Report to the California State Office of Planning.
Williams, G.D.V. (1983), 'Agroclimatic Resource Analyses: An Example Using an Index Derived and Applied in Canada', *Agricultural Meteorology*, 28: 31–47.
Williams, G.D.V., Pock, N.J. and Russwurm, L.H. (1978), 'The Spatial Association of Agroclimatic Resources and Urban Population in Canada', in R.M. Irving (ed.), *Readings in Canadian Geography* (3rd ed), Toronto, Holt, Rinehart and Winston.
Williams, G.D.V. and Pohl, A. (1987), 'Let Them Eat Houses!: The Implications of Urban

Expansion onto Good Farmland', in C. Cocklin, B.E. Smit and T.R.R. Johnston (eds), *Demands on Rural Lands: Planning for Resource Use*, Boulder, Westview Press, 85–96.

Wiseman, C. (1978), 'Selection of Major Planning Issues', Policy Sciences, 9(1): 71–86.

Wong, C. (1976), *Ontario's Changing Population, Volume 1: Patterns and Factors of Change, 1941–1971*, Toronto, Ontario, Regional Planning Branch, Ontario Department of Treasury, Economics and Intergovernmental Affairs.

Wong, S.-Y. (1983), 'Agricultural Change in Canada, 1941–1976', Waterloo, Canada, Department of Geography, University of Waterloo, unpublished PhD thesis.

Yeates, M. (1975), *Main Street: Windsor to Quebec City*, Toronto, Macmillan.

Yeates, M. (1985), *Land in Canada's Urban Heartland*, Ottawa, Ontario, Lands Directorate, Environment Canada, Land Use in Canada Series 27.

Zeimetz, K.A., Dillon, E., Hardy, E.E. and Otte, R.C. (1976), *Dynamics of Land Use in Fast Growth Areas*, Washington DC, Economic Research Service, USDA, Agricultural Economic Report 325.

Zobler, L. (1962), 'An Economic-Historical View of Natural Resource Use and Conservation', *Economic Geography*, 38(3): 189–94.

Index

adandonment of farmland, 10, 28, 79, 94-95, 176
accretionary development, 9, 33, 83, 84, 88, 168
adaptive,
  change on farms, 117-131
  decision making, 14
adoption of innovation, 111
agri-business, 17, 19, 182
agricultural,
  Code of Practice (Ontario), 171
  community, 12, 13
  diversification, 125
  de-intensification, 173, 198
  extensification, 28, 198
  fundamentalism, 22
  intensification, 26, 28, 48, 72, 125, 156, 198
  labour, 71, 102-103
  land-use zones, 167-169
  service network, 78-79
  specialisation, 61, 75, 125
  zoning, 168
Agricultural Land Protection Act, 1978 (Quebec), 161
agricultural (farming) landscapes, 53, 133-134
  categorisation of, 19
Agroclimatic Resource Index, 50, 51
agro-commodity chains, 120
air photo analysis, 31, 32

alternative (ecologically sensitive) agriculture, 18, 19, 72-74, 135
Amsterdam bourgeoisie, summer homes of, 199
Areas of Outstanding Natural Beauty, 181
Auckland, New Zealand, 86, 184
  intensive agricultural production near, 57
Australia, 25, 71, 91, 142, 192
  conservation reserves and national parks, 181
  spatial coincidence of population and productive farmland, 26

biotechnology, 6, 42, 75
Birmingham, UK, 125, 176
Blakely vs Manukau County Council (NZ), 167
Brampton, Ontario, 146
  soliciting farmer opinion in, 180
Brisbane, Australia, 144
Britain,
  debate between farmers and conservationists, 184
  farmland conversion in, 166
British Columbia, 49, 168
  Agricultural Land Commission, 169
  Agricultural Land Reserves (ALR), 169

INDEX

California, 47, 50, 59, 76, 162
   Land Conservation Act (Williamson Act), 145, 161, 174, 175
Canada, 24, 32, 34, 39, 46, 49, 50, 61, 73, 91, 92, 95, 96, 135, 143, 167, 169, 170, 173, 182
   agricultural capability of resource base, 39
   change in agricultural employment, 72
   farmland conversion in, 40, 167
   metropolitan centres, 58
   pick-your-own operations on PEI, 64
   Prairies, 48, 76
   producer marketing boards, 19
   tender fruit production in, 49
   urban growth in, 40
Canada Land Inventory, 38, 39, 51
Canada/US Free Trade Agreement, 14
Canadian Class I Equivalent Productivity Index, 51
capital,
   circuits of, 201
   factors affecting input, 14
   substitution, 6, 17, 91, 193
capitalistic,
   agricultural production, 15, 16, 19, 35, 70, 78
   decision making, 18
   farming, 18, 69
cash cropping on rented land, 127
Charles deGaulle Airport, 107, 180
cheap food policies, 36
Christchurch, New Zealand, 184
Climatic Moisture Index, 50
climatic warming, 48, 49
coefficient d'occupation du sol (COS), 173
Common Agricultural Policy of the European Community, 14
commodification of family labour, 69
comparative advantage, 56
conflicts between farm and non-farm populations, 18
conservation ethic, 19
conservation technologies, 18
cost-price squeeze, 110
Countryside Commission (UK), 176, 182
co-operative agricultural production, 15

cost-influencing public intervention, 174–176, 182

decision making, models of, 114–115
degradation of the soil resource, Canadian Senate Report on, 48
demand for land, 10, 11
direct impact of urbanisation on agriculture, 83
direct marketing of farm products, 62, 63, 125–126, 203
   in Connecticut, 176
disinvestment in agriculture, stages in, 28

ecological considerations, 14
efficiency considerations in allocating resource, 151
England, 84
   farmland conversion in, 27
entrepreneurial,
   activity, 119-123
   behaviour, 133, 197
   decisions, 102, 113, 116, 191, 202
Environment Canada, 33, 34, 50
   Lands Directorate of, 146
equity considerations, 157, 158
   in allocating resources, 151
European Common Market, 150, 197
European Community, 135, 155, 181, 202
   Common Agricultural Policy of, 14
evaluation of public policy, 178–179
exclusive agricultural zones, 163, 168
externalities, 152–156
exurbanites, 13, 79, 90, 97, 157

factory farms, 1
farm firm,
   consolidation (amalgamation), 15, 76, 123, 126, 192
   decision making system, 102–104
   diversification, 67, 127, 176
   equity cycle, 108
   fragmentation, 10, 14
   investment decisions, 202
   planning horizons, 203
   simplification, 129
   specialization, 126
   structure, 11, 72, 76

INDEX

winding-down, 119, 129
farm enterprise structure, 11
   diversification of, 127
   simplification of, 129
farmland
   conservation (preservation/protection) policy, 52, 55, 139, 145, 165, 200, 201
   fragmentation, 88-89
   rental, 12, 14, 92-94, 96, 110, 123, 128, 192, 193, 195
family cycle, 108
family farm, 2, 15, 68, 69, 70, 100, 108, 115, 182, 193
food surpluses, 198
Foodland Guidelines, Ontario, 148-149, 169
   evaluation of, 177-178
foodland perspective, 22
forms of agricultural production, 15
France, 15, 67, 72, 94, 161, 162, 172, 173
   Paris Basin, 15, 47
   food supplies of Paris, 47, 48
   spatial coincidence of population and productive farmland, 24
French and Pickering Creek Trust, 54
Friends of Oregon, 146
fixity of farm assets, 82, 85
functions of agricultural land, 20-22, 139-140, 159, 180-184, 203

Garden City Concept, 183
geographic information systems, 33
geographic scales of analysis, 4, 5, 6, 16
globalisation, 17
goals and values of farmers, 105-109
Golden Horseshoe, 32, 47
good planning principles, 147
government subsidies, 14
Great Britain, 50, 143
   spatial coincidence of population and productive farmland, 24
Greater London Region, 47
Green
   Belts, 84, 164, 172, 182-184
   Wedges, 164, 183

Halton (Ontario) Agricultural Advisory Committee, 180

Hartford, Connecticut,
   pick-your-own operations near, 63
Hawaii, 168
   State Land Use Commission, 179
hobby farming, 18, 19, 70, 72-74, 82, 98, 106, 112, 122, 139, 191, 193, 194, 198
horsiculture, 67
Howard, Edenezer, and the Garden City Concept, 183

Ile-de-France (Paris) region, 31, 53, 98, 176
impacts of urban development, 10, 12, 23
   direct vs indirect, 11
indirect impacts of urbanisation on agriculture, 83-95
industrialised agriculture, 17, 18, 19, 75, 76, 78, 96, 134, 191, 192, 193
   factors encouraging, 19
Industrial Revolution, 1, 24, 150, 199
information, 18
   role in farm decisions, 111-112
inner fringe, 8, 90, 126, 132, 184, 195, 197
innovative decision making, 14
institutional forces, 14
interests in agriculture land, 3, 20-22, 139-141, 159
   collective interests vs private interests, 20-22, 141, 149, 200, 201
internationalisatoin, 182

Jeffersonian Ideal, 51

Kitchener-Walterloo, Ontario, 25, 92

labour markets, 102
land,
   as an input in agricultural production, 79-82, 193
   capability classifications, 50
   conversion, 10, 25-54, 175
   conservation ethic, 199
   speculation, 10
   tenure, 8, 9, 14, 91-96, 98, 100, 102
landscape amenity perspective, 163–164
landscapes of,

INDEX

agricultural adaptation, 19, 187–188, 194
agricultural degeneration, 19, 186–187, 194, 202
agricultural development, 19, 188–189, 194
Land Evaluation Group, 43, 52
Landsat Thematic Mapper, 32, 33
Land Utilisation Surveys, 31
land-use conversion, 10, 25-54
land-use planning, 37, 89, 131, 203
leap-frog development, 92
London's Metropolitan Green Belt, 93, 123-124, 128, 164, 188
Los Algeles, 47
   dairying near, 57
low input farming, 18

market failure, 150-157
Marlborough Sounds region of New Zealand,
   pick-your-own operations in, 64
Maryland Agricultural Land Preservation Foundation, 173
mathematical programming models, 44, 52
mechanisation of agriculture, 1, 6, 74-78
metropolitan or urban-based forces, 9, 10, 11, 13, 116, 186, 195
minimum lot size, 170
Minnesota Metropolitan Agricultural Preserves Act (1980), 174
mixed-tenure farms, 192
modes of production, 15, 18, 19, 20, 70, 96
modification of agricultural communities, 12-13
Montreal, 91, 95, 108, 126, 128, 129, 148
multiple job-holding, 72, 113, 185
multiple functions of agricultural land, 204

National Agricultural Lands Survey (USA), 147
National Capital Commission (Canada), 172
Natural Areas of Regional Importance (France), 183

near-urban agriculture, characteristics of, 190-194
Netherlands the, 6, 21, 162, 182, 199
   Natural Landscape Parks in, 135
New Britain, Connecticut,
   pick-your-own operations near, 63
New Democratic Party (NDP) of Ontario,
   role in farmland preservation policy formulation, 148
New Towns, 164, 165, 189
   around Paris, 9, 180
   expropriation of agricultural land for, 172
New York's Agricultural Districts, 145, 174, 176
New Zealand, 42, 45, 50, 57, 71, 74, 82, 91, 96, 135, 142, 144, 181, 184, 185, 192
   changing census definitions, 30
   farmland conversion in, 27, 40, 167
   pick-your-own operations in Marlborough Sounds, 64
   producer marketing boards, 19
   Rural Downturn in, 113
   spatial coincidence of population and productive farmland, 24
   structural change in pastoral farming, 16
   Valuation Department, 81
Niagara,
   Fruit Belt, 12, 57, 60, 91, 96, 165
   Peninsula, 49, 146
non-agricultural functions of agricultural land, 2, 141, 197, 198, 200
non-farm
   development, 12, 201
   population, 12, 13, 18, 89, 126, 175, 182
non-metropolitan forces, 10, 11, 13, 116, 117, 118, 134, 202
non-traditional farm enterprises, adoption of, 128-129
North America, 26, 31, 54, 57, 63, 64, 71, 93, 94, 98, 123, 142, 145, 156, 161, 163, 167, 169, 172, 175, 192, 198, 200

Oakville, Ontario, 93

INDEX

off-farm employment, 12
Okanagan Valley, British Columbia, 49
Ontario, 24, 32, 44, 49, 50, 60, 73, 91, 127, 146, 147, 148, 149, 162, 169, 170, 171, 175, 177, 189
  Department of Municipal Affairs, 148
  future urban expansion, 42
  Ministry of Agriculture and Food, 130-131
  Select Committee on Land Drainage, 196
opportunity costs, 144
Orange County, California, 58
Oregon Land Conservation and Development Act, 146, 161
organic and alternative agriculture, 19
Ottawa, 13
  Green Belt, 172, 182
outer fringe, 8
ownership of land, importance of, 142

Palmerston North, New Zealand, 144
Paris Region, 9, 15, 34, 47, 48, 57, 60, 67, 72, 73, 76, 86, 89, 96, 107, 125, 128, 129, 130, 133, 161, 180, 183
  Green Belt, 184, 186
  pick-your-own operations near, 64
part-time farming, 18, 19, 70, 72-74, 98, 106, 123, 191, 194, 198
petty commodity production, 16
Philadelphia, 54
pick-your-own operations, 18, 63-68, 126, 130-131,
  in association with farm shops, 125
planners, role in policy formulation, 147
place function, 20, 21-22
play function, 20, 22
pluri-activity, 72
policy decisions on farms, 112-113
policy formulation, 147, 177-178
political-economy perspective, 3
post-industrial society, 3, 5, 6, 10, 11, 16, 18, 19, 56, 63, 71, 72, 120, 130, 182, 191
Pred's behavioural matrix, 111
preferential property tax assessment, 175, 192
  evaluation of, 176

Preservation of Agricultural Lands Society (PALS), 146
preservation of farmland, 135
prime farmland, 40, 41
Prince Edward Island, Canada, pick-your-own operations in, 64
producer marketing boards, 19
production function, 10, 20
property rights, 143-144, 160-161
  transfer of, 172
property taxes, effect of non-farm development on, 90
protection function, 20, 21
public intervention, 55, 68, 84, 90, 101, 134, 136, 137, 138, 139, 141, 143, 149, 150, 157, 158, 200, 202, 203

Quebec, 24, 103, 169, 170
  Agricultural Land Protection Law, 168

Regional Chamber of Agriculture, Ile-de-France Region, 176
Regional City, 4, 6, 7, 9, 10, 18, 71, 82, 204
  internal structure, 7
  stages in its evolution, 9
regional specialisation, 17, 61, 96
renaissance of rural centres, 79
Resources for Tomorrow Conference (1960), 148
resource base for agriculture,
  degradation of, 49, 199
  quality of in the City's Countryside, 201
resource-use
  conflicts, 12, 199
  flexibility and criticality, 52
restructuring of agriculture, 198
right-to-farm, 163, 170
rights in land, regulating, 169-174
Romans' Villas, 199
rural and non-farm elements of the City's Countryside, 10, 204
rural hinterland, 8
rural-urban differences, erosion of, 182
rural-urban fringe, 8, 175
  form and function, 8
  factors affecting geography of, 8-9

satellite imagery, 32, 33
scenario building, 46, 53
Scotland,
  farmland conversion in, 27, 40
Schultz's urban-industrial hypothesis, 111
separation of development rights, 174
set-aside land, 28
shifting economic margins, 28
simplification of natural landscapes, 35
Sinclair,
  anticipation of urban development model, 84–86
  Bryant's elaboration, 85–86
Societe d'Amenagement Foncier et d'Etablissement Rural (SAFER), 172
soil conservation, 201
soil degradation, 23
  Canadian Senate Report on, 48
South-East and West Midlands,
  pick-your-own operations in, 63
Special Committee on Farm Income (Ontario), 91
speciality agricultural lands, 52
St. Catharines (Ontario), 91
strategic decisions on farms, 112–113
Storie Index, 50
structural diversification, 67
sustainable
  agricultural development, 18, 35, 144, 197
  development, 23, 161, 202
Switzerland, 47
  support of mountain agriculture, 54
Sydney, Australia,
  market-oriented agriculture, 57
  metropolitan region, 59
systems of exchange, 3, 4, 10, 13–14, 19, 20, 45, 63, 69, 70, 91, 92, 95, 117, 118, 138, 139, 194
  for farm inputs, 69–95
  for farm output, 55–69
systems perspective, 14

Tax Reform Act, 1976, (US), 175
technological treadmill, 118
technology, 6, 10
  agricultural, 74–78
  communication, 5, 6, 7
  labour-reducing and labour-intensive, 75
  negative impact of, 155
  production, 5, 6, 18
  storage, 61
  transportation, 6, 17, 57, 60
Third World,
  food shortages, 35, 36
  access to resources, in, 37
Thunian analysis, 56, 80–81, 84
Toronto, 67, 73, 79, 89, 91, 93, 107, 108, 109, 112, 127, 128, 131, 146, 180, 189
  pick-your-own operations near, 126
Town and Country Planning Act (UK), 161
Town and Country Planning Act (New Zealand), 167, 170
transport costs, 61
transportation technology, 17
typical changes on farms, 117–119

United Kingdom, 26, 30, 47, 54, 63, 67, 75, 96, 98, 123, 146, 161, 162, 164, 170, 176, 181, 183, 200
United States, 40, 57, 67, 95, 143, 147, 167, 169, 170, 174, 175, 200
  farmland conversion in, 27, 166
  urban encroachment on rural lands, 42
  spatial coincidence of population and productive farmland, 24
University of Guelph, 52
Urban Development in Rural Areas (UDIRA) Policy, 148
urban,
  development pressures, 9
  expansion, 27, 28, 31, 42, 53, 94, 200
  field, 4, 6, 10, 11, 71
  fringe, 11, 33, 58, 89, 90, 92, 125, 174, 176, 179, 184, 193
  sprawl, 166
Urban Growth Boundaries,
  delimitation of, 164
urbanisation of farmland, 25–54
  themes in the debate, 35–36
urban-agricultural relationships, 10
  consequences of interaction, 10–12
urban-industrial complex, 1, 2, 15, 17, 69, 73, 74, 111, 120

Utrecht, 87, 125

Vancouver, 91
voluntary agricultural districts, 174

Western Europe, 21, 26, 54, 63, 64, 71, 135, 146, 156, 198, 200

Wisconsin, farmland preservation policy in, 145

zones naturelles d'equilbre (ZNE), Paris Region, 183